Studying Science
a guide to undergraduate success

Also of interest

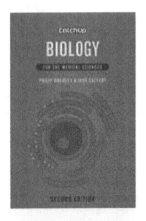

CATCH UP BIOLOGY 2e

P. Bradley and J. Calvert

ISBN 9781904842880

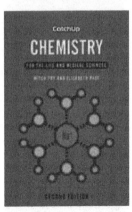

CATCH UP CHEMISTRY 2e

M. Fry and E. Page

ISBN 9781904842897

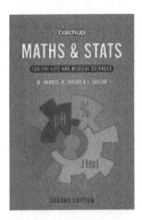

CATCH UP MATHS & STATS 2e

M. Harris, G. Taylor and J. Taylor

ISBN 9781904842903

Studying Science

a guide to undergraduate success

Second Edition

Pauline Millican and John Heritage
Undergraduate School, Faculty of Biological Sciences,
University of Leeds, Leeds, UK

Scion

© **Scion Publishing Limited, 2014**

Second edition first published 2014

First edition (ISBN 978 1 904842 74 3) published 2009

A CIP catalogue record for this book is available from the British Library.

ISBN 978 1 907904 50 9

Scion Publishing Limited

The Old Hayloft, Vantage Business Park, Bloxham Road, Banbury OX16 9UX, UK

www.scionpublishing.com

Important Note from the Publisher

The information contained within this book was obtained by Scion Publishing Ltd from sources believed by us to be reliable. However, while every effort has been made to ensure its accuracy, no responsibility for loss or injury whatsoever occasioned to any person acting or refraining from action as a result of information contained herein can be accepted by the authors or publishers.

Readers are reminded that medicine is a constantly evolving science and while the authors and publishers have ensured that all dosages, applications and practices are based on current indications, there may be specific practices which differ between communities. You should always follow the guidelines laid down by the manufacturers of specific products and the relevant authorities in the country in which you are practising.

Although every effort has been made to ensure that all owners of copyright material have been acknowledged in this publication, we would be pleased to acknowledge in subsequent reprints or editions any omissions brought to our attention.

Registered names, trademarks, etc. used in this book, even when not marked as such, are not to be considered unprotected by law.

Typeset by Phoenix Photosetting, Chatham, Kent, UK

Printed in the UK by 4edge Limited

Contents

Preface to the second edition

Much has happened in Higher Education since we produced the first edition of *Studying Science* and we were delighted to be approached by Scion Publishing to consider updating our book. This has given us the opportunity to review the advice we can offer to students embarking on undergraduate careers in the biological sciences. This job was made much easier by the advice offered by Clare Boomer and Simon Watkins, who were very proactive in suggesting what needed change and what should be kept. We owe them a debt of thanks. On reviewing the old material, it was satisfying how much advice remains pertinent but inevitably where technology interfaces with education, much has changed. Perhaps the most significant development has been the universal adoption of virtual learning environments. These have enriched teaching significantly and we acknowledge this impact with a new section devoted to their use. Our first edition included a detailed 'How to' guide for *Microsoft Word*, *Powerpoint* and *Excel*. With the inexorable rise of video demonstrations available on the internet, and because of the short shelf life of particular software packages, it was decided that this was no longer appropriate. Updating this material every time a new *Office* package is released would be a nightmare. However, experience of marking undergraduate work and even theses presented for higher degrees has shown that many students are unaware of the power of these programs. This edition includes a new appendix where we try to encourage the reader to explore the potential of the *Office* suite and to apply these principles to hone their IT skills. In addition to the people we acknowledged for the first addition, we would also like to thank Chris Reynolds and Veronica Volz for advice on Mac applications. We hope that you find this new material useful.

Pauline Millican and John Heritage
May 2014

Preface to the first edition

This book has a long pedigree. For a number of years, on gaining a place at the Leeds University Medical School, students were given a short "study skills" booklet. This proved very successful and David Bullimore, its creator, very generously allowed his guide to be adapted for new microbiologists. The first of these guides was presented to new undergraduates reading Microbiology in 1995. Each year, feedback from students was actively sought and the guide evolved accordingly. With the merger of the former Department of Microbiology into a larger structure, the focus of the skills guide changed again. It then grew into a Faculty-wide skills guide with major input from colleagues, notably Linda Bonnett, as acknowledged over the page. It also morphed from a booklet to an intranet site. At that point, it was suggested that the material should be expanded to cover help and guidance for success in a life science degree, all the way from gaining a place until graduation.

The University of Leeds has a very large Faculty of Biological Sciences and consequently we see students encountering a very wide range of problems. With a few exceptions, most of these can be minimized by following faithful and familiar remedies. These experiences have contributed substantially to the advice in this book. We have also found that when helping students with problems, it has been very useful to be able to direct them to the Leeds Study Skills Guide. We feel that now is a good time to share our experience and advice with those who are about to embark on a science degree at other universities. That is probably why you are holding this book at the moment. We hope that you will find it as useful as so many of our past students at Leeds have done.

Pauline Millican and John Heritage
Leeds, February 2009

Acknowledgements

Many people have provided help and advice in the preparation of this book. David Bullimore is due particular thanks, for his sound introductory advice given to many cohorts of new students arriving at university over the years; without his generosity, this book could not exist. We would also especially like to thank Linda Bonnett and Peter Millican who gave generously of their ideas and time, and contributed towards several sections of the book; without their assistance, this book would have looked very different. We are grateful to many colleagues: David Adams and Celia Knight for encouraging us to get this project off the ground, Jerry Knapp, Stevie McBurney, Sue Whittle, David Pilbeam, John Grahame, John Walker, Mitch Fry, Linda Wilson and Danni Strauss, for their help during the preparation of the book. There may be others whom we may have inadvertently omitted; it is not that we are not grateful – it is just that this has been such a long project that not everyone's contribution has stuck in our minds! We also want to thank David Hames for his common-sense suggestions and Jonathan Ray for his continued support during the production process. Finally, thanks must be given to Pauline Heritage, for giving her other half the time and space to work.

Chapter 1 Starting out

Welcome to university! This book provides a guide to the skills that you will find of value if you are to get the most from your undergraduate studies. It does not contain any magic formula that will guarantee a first class honours degree, but it does contain common-sense advice to help you to take responsibility for yourself. This advice will allow you to adapt to the challenges of university life, develop good techniques for studying, and really benefit from the opportunities on offer. A university education is about far more than acquiring a body of knowledge – it is about acquiring and practising life skills. The subject-specific, vocational, and 'transferable' skills acquired during university are all vital in equipping modern graduates for successful careers.

Some of the important skills you will need to develop include the following:

- **Personal skills:** such as organizing your work, taking responsibility for your own learning, and learning from others. You must become able to work independently and effectively and practise reviewing your personal development.
- **Managing your learning**: which means being able to reflect on how you learn and being positively self-critical in order to see how you can improve on your own performance.
- **Time management:** being organized and efficient with your time will enable you to be more successful in completing the set tasks, with time left over for relaxation.
- **Retrieval and handling of information:** finding out where to find things out.
- **Problem-solving and critical analysis:** you will find many opportunities to practise solving academic problems during your course. Critical analysis involves evaluation of other people's work, and of your own efforts, in order to decide how reliable or trustworthy various sources of information are. While practising these, you will also enhance your reasoning skills and develop your numeracy (covered in detail in *Catch Up Maths & Stats*).
- **Communication skills:** including producing written, oral and visual presentations.
- **Teamwork:** this is vital for your future career, whatever you decide to do. Often your practical work will involve collaboration with at least one person and sometimes you will work in larger groups.

Developing your ability in the skills listed above will come naturally if you become effective at the tasks and activities normally associated with completing an undergraduate course. Therefore, this guide contains hints and tips on important topics such as

- taking notes in lectures
- contributing to tutorials
- writing essays

- making the most of practical classes
- summarizing/extracting information
- using libraries effectively
- revision skills
- coping with examinations, in particular how to answer questions appropriately

1.1 The transition to university life

You probably have mixed feelings about the start of your undergraduate career. You might feel proud to be a part of *my* university, excited about the course you will study, looking forward to meeting new friends, and to enjoying a wide range of cultural experiences. You may also feel very daunted at the prospect of leaving home to become independent for the first time in your life; being in charge of your own finances, accommodation and catering – all of this can be scary at first. You may believe that everyone else knows so much more than you do about almost everything. But remember – it is very likely that they are experiencing exactly the same emotions as you!

More mature students and those with non-standard routes to higher education can feel just as daunted as those coming straight from school. Whatever route you have taken to get here, however, the most important thing to remember is: **DON'T PANIC**. Your worries are perfectly normal, and the best way to overcome them is to turn them to your advantage. Just think about all the interesting people you will meet and from whom you can learn much.

Whatever your background, it is probable that most of the teaching that you have received has been delivered in small groups, probably less than 30 of you, by a teacher who knew you, and whom you came to know well too. Although you may receive some small group teaching at university, for much of the time you are likely to be taught in lectures, possibly with hundreds of other students. You may have never seen the lecturer before, and you may never come across that particular individual again. This may be disorienting at first, but you will find that you get used to it very quickly.

One of the greatest privileges of a university education is that you have the opportunity to learn about far more than just the subject you came to study. You will mix with and make friends with students studying diverse disciplines and you will have many opportunities to join a variety of societies and clubs. Taking an active part in such social activities is all part of discovering about the world, and you should bear in mind the old saying *"all work and no play…"*.

Below is a short essay entitled "Fresher pressure" adapted from an original by Hannah Jones, a Leeds Biochemistry graduate. It says it all!

> *There can be few more stomach-churningly terrifying ordeals than coming to University. Leaving home is your final break from childhood, and nothing prepares you for it. As you begin to embrace the horrors of launderettes, supermarket shopping and using an oven, you experience pangs of fondness for your parents which you may never have imagined possible.*

Years later, as you and your bosom buddies will pack up your posters and fight over who keeps the ironing board, once used as a tennis net, you'll look back on your early student days and laugh as you remember those crippling nerves.

But when you first leave your beloved homestead for the big smoke, it can be hard to think rationally about the coming months. Everyone will have offered you well-meaning words of wisdom about those early weeks, but don't believe a word of it. Let me dispel at least one myth...

... You'll think that you will never find any friends like the ones to whom you bid a fond farewell in the days before your departure. You decide that you'll never meet anyone who shares your love of handball and the "nonny no" madrigal, and that your new flatmates will make a joint decision that they all hate you, sneaking away behind your back for marvellous nights out while you're in the shower or engaged in a stiff upper lip telephone conversation with your mother.

This is pure, but understandable, paranoia on your part – as you will realize about three weeks into term time as you stagger home from a club singing "Take Me Home Country Roads" with all your other mates from the Handball Society. Even though the University community can seem the size of your entire home county, it does mean that you never have too far to go to find someone with similar interests to yourself.

The one true thing you're likely to be told about University is that you will meet your friends for life during these few years. Though it may take a time to find them – never fear, you will!

1.2 How degrees are awarded

Traditionally, degrees in UK universities were awarded after a period of study, with the classification of degrees being dependent on a single set of examinations – finals – although some institutions did vary slightly from this pattern, with performance outside finals contributing a small amount to the final degree classification. Typically, success rested on a student's performance in the two or three weeks of these finals examinations. Any mistakes made would be reflected in the class of degree that was awarded. While this model served most people well, it could have devastating consequences for a small minority.

The time taken to complete a degree varies. Within the UK, the honours degree programmes in universities in England, Wales and Northern Ireland typically take three years, unless students opt for a fourth year to develop their professional skills in industry or elsewhere, or in studying abroad. Scottish institutions, however, have always been different because the secondary school system never included "A" levels, having a broader approach that culminated in the award of Highers. This means that students in Scotland moving to their university have a broader base on which to build their tertiary education but, it could be argued, they lack the depth of training fostered by "A" levels used in the rest of the UK (and elsewhere). An important consequence of this is that, traditionally, Scottish honours degree programmes are four years long. Another significant difference between degrees from Scottish universities and these from elsewhere is that the fourth year of study in a Scottish university is the "honours" year. While most students take this route, a few exit after three years with an "ordinary" degree. An ordinary degree from a Scottish university is thus an honourable exit – for other UK universities, people who are awarded ordinary degrees have not made sufficient progress to achieve a degree with honours. Candidates with suitable qualifications from elsewhere may be able to negotiate direct entry into Level 2 of selected degree programmes in Scottish universities.

1.2.1 Modules

In recent years, higher education has undergone a revolution, particularly in the way degree subjects have been taught and examined and "modularization" has been introduced. In modular degrees, programmes are divided into bite-sized chunks (modules) that are delivered and examined largely independently of one another. This system fits some programmes better than it does others and in science it is probably not desirable to treat all information as discrete packets. Consequently, as you progress through a particular programme of study you will find that programmes at higher levels will require that you have taken and usually passed certain "pre-requisite" modules from lower levels.

Each module is allocated a certain number of credits, which reflect the amount of work required to complete that module. Typically, at all levels, the academic year is divided into modules worth a total of 120 credits – occasionally, programmes may have more credits, particularly in those programmes that run field courses in vacations. Different universities use different weightings to modules and there is a wide variety of credit ratings.

1.2.2 Semesters

Along with modularization, most UK universities adopted "semesterization" – splitting of the academic year into halves, or "semesters". Typically, students take modules worth 60 credits in each semester, although individual programmes may allow some flexibility in the number of credits taken in each semester. The academic year may be front-loaded, with students taking 70 credits in the first semester, leaving just 50 credits to be taken in the second semester. Alternatively, the year may be back-loaded, with 50 credits being taken in the first semester and 70 credits being taken in the second semester.

1.2.3 Credits

Throughout your degree, you will be required to gain a certain number of credits by passing specific modules. Even a student who scrapes a bare pass is awarded all of the credits for that module. The actual number of marks achieved in each module is usually derived from both in-course assessment and end-of-module examinations.

The relative weightings of these will vary between modules but this information will be published, so that you can monitor your progress through the degree scheme.

1.2.4 Degree classification

Universities differ in how they calculate the final degree classes but typically only results from study at Levels 2 and 3 contribute to the final degree classification. At Level 1, generally all that is required is for students to pass the requisite number of modules to progress. You should check for yourself exactly what your university does.

Most students graduate with honours. Those who fail a small number of modules at either Level 2 or 3 may have successfully completed sufficient modules to be awarded ordinary degrees. Where there have been significant failures and even an ordinary degree is beyond the grasp of students, they may be eligible for the award of a Diploma or a Certificate of Higher Education. Diplomas are often awarded to those who have sufficient achievement at Level 2; Certificates are for those whose achievement is confined to Level 1 work.

Honours degrees are divided into different classes: first, upper second, lower second and third class. An upper second class degree may be referred to as a 2(i) and a lower second class degree as a 2(ii). While many students aspire to first class honours, the crucial division is between upper and lower second class honours degrees. This is because to qualify for registration for a doctoral programme, an upper second class degree is the cut-off for support from most sources of funding.

Degrees that are classified as 2(ii)s are also known affectionately as "Desmond's" after Desmond Tutu, the Archbishop of Cape Town. Third class degrees may be referred to as "Richard's" (after Richard III) and in the 1980s and early 1990s were known as "Douglas's", after Douglas Hurd, the Home Secretary then Foreign Secretary at the time.

The separation of UK degrees into classes has a long history but is different from classifications used in other parts of the world. For a long time, individual institutions have been working on an alternative system, with the "Grade Point Average" being the most likely alternative to emerge, should changes ever be made.

1.2.5 University standards

To ensure that the quality of degrees is maintained and that the standard of degrees is similar across universities, there are several checks in the system. The most important is the appointment of external examiners – senior academics with broad and deep experience of teaching and assessment. It is their job to approve degrees and to report to the university on the quality of the examination process. External examiners highlight any issues that may require attention and provide advice on new modules and programmes. Additionally, universities have internal self-review schemes and they are periodically reviewed by external assessors.

The aims of your degree programme are to provide you with a wide knowledge of your subject, to give you an appreciation of its latest research and future direction, and to prepare you for your future career.

Chapter 2 Your degree – preparation for a career

What influenced your decision to embark on a degree course? Was it because:

- you felt that you need a degree in order to get a good job?
- you enjoy the subject and want to discover more?
- you wanted to get away from home and become independent?
- all your friends were going away to university and you didn't want to be left behind?
- you wanted to experience student life?

You may feel that all of these apply to you, but it is not really necessary to go to university to leave home and gain your independence, or to find a job where you work and have fun with people of a similar age. Taking a degree course involves a lot of time, hard work and, inevitably at times, a great deal of anxiety. In addition, there will be the strong possibility that by the end of it you will be in considerable debt.

Now ask yourself: what do I want to *get out* of being at university? Most students would answer:

- I want to get the best degree I can to help me get the best job that suits me, as soon as possible after I leave university
- I want to study a subject I enjoy

So the reasons that you embark on a university degree course in the first place become clearer; you want to spend your time doing something in which you are interested to help you get **the best degree result you can in order to help you have an enjoyable and rewarding career.**

How can you go about this?

The following sections in this book contain lots of information about how to work effectively and successfully at university and, if you follow the advice, you will put yourself in a very good position to do well. Doing well may not mean doing *as well* as you could, however! It would be a great shame if, having spent years of your life enjoying the privilege of being in higher education, at the end you achieve a less than satisfactory result. This section is devoted to **strategies to help you achieve to the VERY BEST of your abilities** while at university.

2.1 Skills for success

At university, you will have a lot more freedom to plan and structure your own work than at school or college, but you must not be tempted to socialize at the expense of your subject work. If you do, you will quickly find that you will start to fail. It is important early on that you try to get the balance right between work and socializing, to allow you quickly to establish a solid foundation on which to build your academic career. Adjusting to this self-discipline can be hard, as can be trying to understand exactly what is expected. If you find that this is the case, then do not be afraid to ask your tutors or fellow students for help and advice.

Also, bear in mind that, although at school or college you may have been able to improve the first mark you obtained in an exam or assessment by retaking it, at university this is not normally possible and your first attempt at an assessment is the one that counts. If you are allowed to re-sit a failed exam, the mark you can obtain may be capped at just the pass mark. This means that your first attempt at the exam really should be your best.

At Level 1 (your first year), the foundations are laid for your course; you will gain an overview of your subject at an elementary level and begin to develop the skills necessary to progress. These will be both subject-specific (e.g. experimental design, laboratory techniques, data presentation, etc.) and also generic (e.g. IT, time management, communication, problem-solving). By the end of this year you will be expected to know the core aspects of your subject and its principal conventions and terminology. The skills you have acquired, such as writing essays, carrying out practical exercises and producing reports, should enable you to demonstrate your knowledge and expertise in your subject.

Level 2, or second year, is where you develop your learning capabilities and explore your subject. Your knowledge and understanding will become both broader and more detailed and you will develop more independence. The subject-specific skills that you acquire are required to carry out more demanding and complex tasks as you progress.

Level 3, or third year (even if you have a study year abroad or in industry), is the time for you to consolidate your knowledge and expertise in the subject. You will be expected to explore trends in current research and be aware of possible future developments. Generally, this is the year in which you will, independently, carry out an individual research project; a challenging task but one that will help you become proficient at many of the skills you practised in the previous two years. This will give you a better view of what it is like to be a research scientist.

2.1.1 *Planning your study and leisure time*

OK, there's a lot going on at university and it's a shame to miss all the other activities that are part of it. How do you study effectively and still have time to do other stuff?

Good planning is the key to success

Ask yourself "How do I plan my study time?" Is it weekly, monthly, termly or even yearly? All answers could be correct depending on what you are doing. Successful students plan by a combination of all of these, using the week as the basic unit for essays, reports, worksheets etc., but bearing in mind events and deadlines on a longer timescale, such as examinations and projects. Ideally, you should:

- keep a diary and/or compile a weekly chart to use as a timetable, such as the one shown below (Figure 1.1). Why not produce your own template? You could easily create your own template using the facilities in *Word* or *Excel*, or alternatively you could maintain an online calendar such as that available in Google.
- enter all your lectures, practical classes and tutorials into your diary and/or onto the grid to help you to plan when you are going to fit in the time for your private study.
- try to arrange your study at times when you are most receptive to learning. Are you a lark or an owl? Some people work best early in the day (the larks), whereas others are most receptive at night (the owls).
- do **not** try to burn the candle at both ends on a regular basis; it becomes counter-productive *very quickly*.
- if you think you need active reminders of forthcoming deadlines or meetings, why not set reminders, for example using your smartphone or from an online calendar?

	Weekly Plan – Week Beginning: __/__/__						
	Mon	Tue	Wed	Thu	Fri	Sat	Sun
06.00–08.00							
08.00–09.00							
09.00–10.00							
10.00–11.00							
11.00–12.00							
12.00–13.00							
13.00–14.00							
14.00–15.00							
15.00–16.00							
16.00–17.00							
17.00–18.00							
18.00–19.00							
19.00–20.00							
20.00–21.00							
21.00–22.00							
22.00–23.00							
23.00–24.00							
24.00–01.00							

Figure 1.1 Timetable for study and leisure

2.1.2 *How much work should I do?*

The amount of work that you do from week to week will vary, sometimes quite considerably. On average, however, you should find that your total fixed work commitments for the week should be about 40–45 hours. If you find that you *regularly* need to work for more than 50 hours a week, then you should discuss your work patterns with somebody: friends or classmates, your personal tutor, or another member of staff. You may need advice on learning techniques, or you may have unrealistic ideas about what is actually expected. To help you to get your balance of work correct, you might like to include planned leisure activities, such as sport, society events, or just meeting up with friends, on the timetable as well. These will give you times in the week to which you can look forward.

Use planning timetables sensibly and remember that they should only be used for guidance *not* as rigid timetables that dominate your life. Being flexible is necessary when you have several assignments due close together and you will have to devote considerably more time to your studies than at other times. Remember, too, that this may happen at the end of term when there is considerable pressure to socialize. To make the best use of your time, you should review the *actual* time that you spend doing things, and compare this with your *planned* timetable. Essays, for example, frequently take far longer to prepare and write than you might anticipate. Reviewing your progress will help you to get the balance right when planning future timetables. Besides planning your weekly schedule, you might like to consider using a chart for the whole year. This will help you to avoid being caught out by deadlines for assessments and examinations that otherwise have a nasty habit of creeping up on you. You should plan for your leisure, too: socializing is very important and it can be a great incentive to get work out of the way and then to let your hair down and celebrate.

2.1.3 *Effective private study is an essential part of your course because …*

- it helps you to reinforce the information and to understand the concepts presented to you in lectures and practical classes and to prepare for assessed in-course tests
- regular private study will reduce the pressure on you when it comes to the exams, as you will have done much of the groundwork
- it will increase your understanding and, hopefully, your enjoyment of your subject
- the ability to structure your own time effectively is an invaluable life skill

More detailed advice is given later on how to make the most of the teaching sessions at university but there are two "golden rules" to remember for effective private study.

2.1.4 *Golden rule 1 – plan your time*

Set aside some time for study and then get on with it. Remember, "*procrastination is the thief of time*". If your room is not conducive to study because it is noisy or your friends keep calling round for coffee then go to the library.

2.1.5 Golden rule 2 – use the time effectively

Here are a few tips to help you to achieve this in lectures, tutorials, and practicals.

Lectures

- After each lecture, you should read through the notes/handout. Make a summary. If there's something you don't understand, look it up in one of the recommended textbooks. If you still don't understand, ask for help. Stupid questions are those that don't get asked.

- Repeat the process at the end of a lecture block. Can you summarize the key points of this set of lectures? Summarizing what you know makes an excellent resource for revision.

- Make use of resources on the internet – but make sure you can trust the sources of the information that you use.

Tutorials

- Be prepared for tutorials – complete any preparation work beforehand and hand in, **on time**, any homework for the next session.

Practicals

- Try to read through the relevant notes before you go to the lab session. You may not understand everything at this point but it will give you some idea of what the practical is about and this will increase your understanding and the amount of benefit you get from it. You will also **get it done more quickly**.

- Write up practical reports and **hand them in on time**. A poor/incomplete record of practical work is the best way to fail the module.

2.1.6 And finally... the responsibility for learning rests on you

Lecturers and demonstrators are anxious to help you and want you to enjoy the course, but can do little if you are not prepared to make the effort. Interest often comes through understanding.

2.2 Help!

Things do go wrong. When you start your undergraduate course, you are embarking upon a new life. Adjusting to a new subject and a new way of studying is not easy. It can be daunting and many students find it difficult at the outset. Don't be discouraged – ask for help. Remember that...

> *...asking for help is not a sign of weakness, it is good common sense.*

Most universities recognize that it is reassuring to students to have a member of staff whom they get to know well. Most universities run a **personal tutor system** where the tutor will meet students regularly, particularly during the early part of your course. This enables students to get to know their personal tutor quickly and so should be able to turn to him/her if general difficulties arise. Alternatively, you may feel happier talking to another member of staff. It is usually easy to spot the people who will be sympathetic to this approach.

2.2.1 *Where else can you ask for help?*

If you are having difficulties with your studies a fellow student may be an appropriate source of support, but it may be better to talk to a member of the teaching staff. You can always take your academic problems to the person running the module. On a personal level, it may be that a chat with a friend will sort things out. A quick phone call home can work wonders, too.

In every academic year, many students suffer from 'flu-like illnesses, as people from all over the world arrive at university in the autumn, bringing with them their "local" viruses. This will almost inevitably lead to absences from some classes. Do not forget to copy up the notes of lectures that you have missed from a friend who is willing to lend you a *good* set of notes. Even if you have access to a full set of notes, it is worth checking whether any additional information was given in a session that you have missed. Your tutors will be sympathetic to *genuine cases* and will expect you to report absences as soon as possible after your return, or by phoning when you are still away. This is particularly important if you are likely to be off for more than a couple of days.

Do remember to register with a local medical practice when you arrive at university. Although you may wish to register with a local General Practitioner (GP), most universities have a student medical practice that offers the full range of general practice services, and provides the same high degree of confidentiality as your GP at home. In addition, the doctors in university practices have a very wide experience of dealing with "student" problems.

Dear John... the agony aunt page

Dear John,

I have got a "friend" who seems to be losing interest in the course. He is really bright but all he does is slob around the flat watching daytime telly. He exists on takeaways and cans of beer. I know that he is going to fail the year if he carries on like this. What can I tell him he should do?

Worried of Woodhouse

Uncle John's advice:

When your "friend" starts getting turned off it is *REALLY IMPORTANT* that he should see his personal tutor or *someone* on the academic staff *as soon as possible*. It may be that he just needs to chat about his situation and about life in general. Find a person who can help your "friend" to make the best decision for **him** about **his future.**

If your "friend" stops attending and gives up on course work then some pretty unpleasant things will start to happen. Your "friend" will get letters sent to him, inviting him for a chat (i.e. he will have to explain his behaviour). If your "friend" ignores these approaches, he will get letters from the university authorities. If, foolishly, your "friend" ignores these, or fails to convince the powers that be that he really has tried hard, then he will be kicked out. Imagine how his parents will feel about the support and money they have invested!

What your "friend" must do is talk to someone before the problem becomes a tragedy. Remember, in the words of H. G. Wells, "... *failure is not failure nor waste wasted if it sweeps away illusion and lights the road to a plan*".

2.3 Personal Development Portfolios

Many university departments encourage their students to maintain a Personal Development Portfolio (PDP). As its name suggests, the aim of a PDP is to encourage students to get into the habit of thinking about and keeping a record of their personal development so that they become more aware of their own strengths and weaknesses. This means reflecting on performance in all aspects of both academic progression and also personal skills (such as the ability to communicate well, to present information clearly and in an appropriate form, and to work well in a team), recognizing strengths so that they can be built on, and identifying weak areas so that efforts can be made to improve in them.

2.3.1 Why should you keep a PDP?

If you take a reflective attitude to your personal development you will be taking responsibility for your learning, and you are more likely to take full advantage of the opportunities available at university for developing your knowledge and skills. This will achieve one of the aims of higher education – to become an independent learner. You will also be much more likely to reach your potential, because only *you* can truly identify the weaknesses that need to be addressed, and only *you* are truly able to organize your time to set about improving those areas. Being in control of your own learning will make you more aware of your abilities, more self-reliant, and also more capable; attributes that most employers will be looking for in prospective graduate employees. In addition, when it comes to writing your *curriculum vitae*, or CV for short, a PDP will be invaluable for reminding you of your achievements and skills.

To get a good idea of the types of skills and attributes that are valued by employers, you only have to look through adverts for jobs. A report published in 2008 entitled '*Graduate Employability: What do employers think and want?*' by Will Archer and Jess Davison includes a list of attributes that employers most want in new graduates. An adapted version of the top ten is reproduced below:

- communication skills (written and oral)
- team-working skills (and cooperation skills)
- integrity (honesty, dependability and reliability)

- intellectual ability
- confidence (enthusiasm and energy)
- character/personality (self-motivation, commitment, a willingness to learn and a 'can-do' attitude)
- planning and organizational skills (including time management, punctuality, and the ability to meet deadlines)
- literacy (having good writing skills and being able to assimilate written information with ease)
- numeracy (being good with numbers)
- analysis and decision-making skills

When developing your PDP, you should consider how you match up to these criteria. If your university has not organized a PDP scheme for the students, don't let that stop you having a go on your own. The advice below will help you take the initiative for your personal development.

2.3.2 How to keep a PDP

Start thinking strategically:

- identify what you want to achieve and find out what you have to do to achieve it
- evaluate your progress regularly (after a set time: a term, a semester, a year) to see whether you are on course to achieve what you want
- be honest with yourself and do not compromise
- if you are not satisfied with your progress in a particular area, look for resources to help and be prepared to spend extra time on developing your skills in that area
- don't procrastinate – just do it!

This process requires discipline and you will need to keep a log of your progress in such areas as:

- attendance at lectures, tutorials and practicals (is it sufficient?)
- preparation for tutorials and practicals (how much do you do and is it sufficient?)
- participation in tutorials and practicals (how much do you do and is it sufficient?)
- performance in tutorials and practicals (in-course marks for essays, worksheets and practical reports – are they satisfactory?)
- academic performance: exam results (are they satisfactory?)

You could also focus on specific aspects of these areas for development, such as those required to conduct practicals successfully, as listed in the grid below (Figure 2.1). If you feel that you are not progressing satisfactorily, you will need to decide what you need to do to help yourself improve. Tabulating the log of your development, in a form similar to the grid below, will make it easy to identify areas on which you need to concentrate.

Area for development		1 = very poor, 5 = very good				
		1	2	3	4	5
Lectures	preparation					
	attendance					
	follow-up work					
Tutorials	preparation					
	attendance					
	participation					
	tutorial marks (essays, worksheets)					
Practicals	preparation					
	attendance					
	understanding of practical techniques					
	working with a partner or within a team					
	organizational skills					
	time management					
	quality of practical results					
	analysis of results					
	practical marks					
Exam results						

Figure 2.1 Suggestion for a development log

In addition, you should keep a record of the extracurricular activities in which you have taken part. Being the chair, secretary or treasurer of a university club or society shows that you are a reliable and responsible person, in addition to having a life 'outside' your chosen subject. This is especially important if you go into a job unrelated to your subject of study. You should ALWAYS keep a record of all your examination results. Now let's explore how to get the most from your degree programme.

Chapter 3 How to learn in a university setting

3.1 The learning process: "deep" and "surface" learning

Which method of learning do **you** use?

Method A	Method B
"I get down most of the bullet points from the *PowerPoint* presentations and I try to remember those"	"I try to find the main principles underlying new knowledge and try to fit them into what else I know"
"I find cramming for exams suits me. I only need to remember the things that I've learned the day before the exam. Otherwise studying just gets in the way of my social life"	"I try to find practical examples for new material. This can be difficult, but these flashes of understanding come when it all drops into place. Then it just becomes part of what you **know**"
"I collect loads of facts, and keep on writing them down. Then, in the exams, I fit the facts to the questions and hope for the best"	

"... and if you could use only one, which would you choose?"

Surface learning is an approach to learning in which the detail of information in a text or lecture is memorized and is typified by Method A.

Deep learning is characterized by placing the focus on the overall structure of information, and the way that it fits into and extends an existing framework of knowledge and understanding. It looks at a big picture, and fits more knowledge to it to make an even bigger picture, as described in Method B.

Either approach is likely to lead to reasonable results until you start as an undergraduate. *From now on, however, surface learning is less and less likely to lead to success.* Sometimes, a shallow approach to learning may be OK, but the volume and complexity of the new material that you will come across means that, to be successful, you will need to take an in-depth approach to your studies.

To engage in deep learning involves some effort, particularly if you are not familiar with the methods employed. **Persevere with it!** The more you practise deep learning, the easier it becomes.

Independent research has shown that *without exception* students whose work was of a sufficient standard to earn a first or upper-second class honours degree used a deep approach to learning. Students who are awarded third class honours all use surface learning. What more can we say?

3.2 Individual study – how do students learn?

Individual study will certainly form a major part of your learning at university. Because of this, you will need to find the answers to questions such as "where?", "when?" and "for how long?" do you need to study on your own. This section is devoted to individual study, how to go about it and the potential problems that you may encounter.

3.2.1 *Where to study?*

You need to find places to work in which you feel comfortable and, importantly, where you are able to concentrate. You must also be able to have your books and other study materials around you such as in a study-bedroom, a flat, or in one of the libraries. If you use your own room, set it up to suit your needs, with adequate light and heating, and your books, files, etc. close to hand. Many people find that it helps to have one area devoted to study. It is also a good idea to separate the area in which you study from the area in which you sleep. You need quality rest, particularly when working very hard, for example, in the run-up to examinations. Enjoy the time that you spend in social areas such as coffee bars, bars, and the student common rooms, but *never fool yourself that these are places for serious study*.

3.2.2 *When to study?*

The question of when to work is best defined by your experience of your own working pattern. Having studied hard for university entrance, you will probably know at what times you work best. Make these times your prime times, and make sure that you use at least some of your prime time on most days for a concentrated period of study without any intrusion of other activities.

Whatever time of the day or night that you work, you must ensure that you get enough sleep. If you regularly work from six to eight in the morning that is fine, but it is a terrible idea if you regularly stay up until past midnight as well.

3.2.3 *For how long should you study?*

Learning is an active process and few people can really concentrate for more than an hour at a time without a break. *Plan* short five- or ten-minute breaks in your study, every so often, where you have a cup of coffee or tea, talk to your pet piranha, or go out for a breath of fresh air. These breaks will renew your concentration, refresh you, and make you more receptive to new material.

Plan how long you intend to spend at a particular study session, and try roughly to stick to it. You may find it useful to use the timetable approach outlined earlier. This is particularly useful in the first few weeks of the course, when you need to establish your new work pattern. Fill in your fixed commitments, and then hatch in the additional hours that you intend to study.

Be realistic

If the planning approach is going to succeed, then you must be honest about how long you really need to spend studying in order to make sure that you are not being left behind. But also remember that you should not spend every waking hour studying to the exclusion of everything else; a university education should be about much more than obtaining a degree in your chosen subject.

3.2.4 Should you study by topic or by time?

You should try to study by topic, rather than for a certain time. Don't aim to study for *"two hours"*, but aim *"to understand and be able to draw out the Krebs cycle"*. This topic study approach should be an active approach with realistic goals. Write down what you want to start with, list the main points that you want to learn, then sketch in the more detailed bits. You can divide the knowledge into levels: "must know", "should know" and "nice to know". For the Krebs cycle it might look something like the table below.

Table 3.1 "Must know", "should know" and "nice to know" information

"Must know"	"Should know"	"Nice to know"
The framework of knowledge to understand the topic and proceed further. "Must know" may include information such as "… the Krebs cycle is a major energy-yielding pathway. When the Krebs cycle doesn't work, you end up getting cramps". This is "must know" for all serious students of biological sciences, but probably not all that you will need to know about this topic.	Knowledge that adds breadth to your learning, and will help to bring topics together. For example, "should know" information may be "… the Krebs cycle is also known as the citric acid cycle, or the tricarboxylic acid or TCA cycle". This will help you look up alternatives in book indices where only one name may be used. If you are a biochemist, "must know" information will also include "…the enzymes involved and the metabolic components of the Krebs cycle: citrate, *iso*-citrate, α-ketoglutarate, succinyl CoA, etc". For the rest of us mere mortals, this is probably just "should know" or even "nice to know" information.	Knowledge that may be an anecdote or a piece of history. Who *was* Hans Krebs anyway? You can easily survive without "nice to know" knowledge, but it comes in very handy for trivia quizzes.

When you are close to the end of an individual study session, take a few minutes to review what you have covered. Try to tie it in with what you already know, and with other things that you have been studying recently. This will help you to build up an overall picture of your subject, and will greatly improve your understanding. *The more that you practise this, the easier it gets*.

3.3 Effective reading

You will undoubtedly need to spend a large amount of time 'reading for your degree'. The sheer bulk of reading matter to which you might be directed during your courses can be overwhelming. This will come in a variety of guises, including Recommended Reading, Key References, Reading Lists, Directed Reading Lists, or General References. **You will not have time to read all of these,** nor will you be expected so to do. Much of the material that has been suggested will be of great value to you if *you* wish to pursue a particular area in depth. By employing suitable strategies, however, you will become effective in selecting the most appropriate reading material, and decide upon the appropriate amount of detail, breadth, and depth you need. If you are unsure about what is expected of you, then always make sure that you ask what is required.

To read *effectively*, you have to be clear about your purpose. Ask yourself ...**"Why am I reading this?"** To practise, ask yourself now why you are reading this book. Are you just skimming through the text to get a "feel" for the topic, or are you reading it carefully to gain detailed information that you intend to use? Depending upon your answer to this question, your approach will be quite different. To help with serious reading, you may wish to consider the approach that is summarized by the "SQR⁴" approach:

Scan	–	read through for an overview
Question	–	decide what you want from your reading
Read	–	an *active* process, searching out key material
Recall	–	periodically consider what you have covered
Review	–	at the end, pull all the information together
Relate	–	tie in what you have read with other topics

For deep learning, relating what you have read to what you already know is the most important of effective reading disciplines.

Some things are implicit in this approach.

- You should read a definite topic and not just "the next chapter" or "for half an hour" (most textbooks are not designed to be read slavishly from cover to cover).
- Try to familiarize yourself with the overall structure and content of an unfamiliar book, so that you get an overview of the subject.
- You should try to fit new material into a framework of your previous knowledge. This is how new material is most easily learned.

Academic reading is an active rather than a passive process. It is *hard work*. So, you should remember that you will read best somewhere that you can concentrate and work in comfort. Note-taking is a skill that will not only help you in lectures (see *Section 4.1*) but also in your private study. If you learn best by underlining or **highlighting** key sections of text as you read them, do not forget the **Recall**, **Review** and **Relate** sections of the **SQR**[4]. Remember that you must **NEVER** mark library books in this manner; only mark your own private copies of books. If you got a great deal from a book that you have borrowed, consider buying a copy for future reference.

Checklist for effective reading

- Ask yourself (question) what you want to get out of your reading.
- Scan through the chapter(s) to see what the sub-headings are about.
- If the chapter has a summary, read through this more carefully, identifying the main points of the arguments that have been covered.
- Read through the introductory paragraphs carefully; these should set the scene for what is to follow.
- Select the appropriate sections and sub-sections of relevance and actively read these. Identify the main ideas in each section. These are often encapsulated in the first line of a paragraph or section, although sometimes they may be found in the last sentence.
- Study the examples that are given. Make sure that you understand how they relate to the idea of that section of text.
- Summarize periodically the main ideas and examples that you have read in your own words so that you can understand the topic. Making brief notes will help with this. Do not attempt to copy out the textbook! Simple pictures and diagrams can help to give a different slant on a topic and will help you to recall the material. Check that you have not missed anything out and that you really understand the material.
- When you have finished reading the whole topic, review what you have read. The easiest way to do this is to scan through your notes.
- Relate what you have just read to other areas of knowledge.

3.4 When the good times go bad

All this advice on how to learn sounds easy – a breeze, in fact. But there will be bad times when you don't feel like working. Nothing seems to go in, and suddenly you can't see the point of it all. Everything becomes too much bother. Being honest, these times will happen to everybody at some point. The trick is not to let the bad times last for too long. There are no magic solutions to overcoming the bad times, but here are some suggestions.

If you have difficulty in starting your work, then you should choose something constructive but straightforward. For example, you could just note down the main points of a lecture or tutorial. If you find one of the recommended textbooks hard to understand, then try looking at other books, they may have clearer explanations, or they may have better illustrations.

If you are bored with what you are studying, try dividing the topic that bores you into manageable packages, and spend time, little and often, on these smaller sections. Alternatively, you could ask a friend to teach you the topic "in the style of your favourite lecturer". Explaining topics to one another in a fun manner is an excellent way for you both to learn. Finally, you could try mild shock therapy by attempting questions off old examination papers under examination conditions. However, bear in mind that courses change with time, so you may not have the background to answer all or any of the questions on older exam papers. If you have any such problems, consult the person who runs the module about the nature of the course in the past.

None of this may work. Remember that you came to university to train as a good scientist. If things seem to be getting on top of you, discuss your problems with somebody. Members of staff are there to help you. Don't forget your friends. They will be happy to help you out, too.

Chapter 4 Making the most of teaching

4.1 Blended learning and virtual learning environments

Imagine that your timetables show that you are about to have a lecture on *"mycobacterial infections"*. A few years ago, the only way to get a real grip on this topic would be to wade through textbooks, some of which may not even have any relevant pictures. Today you will be able to log into a **virtual learning environment** (VLE) or **portal**, where the lecture course will be backed up with a variety of other resources.

As well as a full set of notes, your tutor may have provided links to a range of additional, complementary resources. These could lead you to a very diverse range of related topics, with one link taking you to Dirk Bouts' painting of Christ in the House of Simon the Leper (*circa* 1445) to illustrate how lepers have been and, maybe, still are regarded as outcasts by most people. The next link may be to a Shakespeare folio for Hamlet; his father's ghost said this of his murder:

> *'... Upon my secure hour thy uncle stole*
> *With juice of cursed hebona in a vial,*
> *And in the porches of my ears did pour*
> *the leperous distilment...'*

Other links may take you to the World Health Organization's data on global trends in the incidence of leprosy, scrofula, the Science Museum's etching showing King Charles II dispensing the "King's Cure" and then on to Regency dandies with high collars, intended to hide their scrofulous necks.

A different direction could focus initially on Henry Purcell, whose early death is thought by many to be due to tuberculosis. Further links via Purcell's music for the funeral of Queen Mary, may lead you to a video clip from Stanley Kubrick's "*A Clockwork Orange*", with its adaptation of Purcell's music played on a synthesizer. An alternative explanation for Purcell's untimely death, that he was poisoned after eating a bad batch of chocolate, may trigger a link to another lecture in the current module devoted to food poisoning. Further links from TB-related deaths could lead you to: the last act of Verdi's "*La Traviata*" in which the heroine, Violetta Valerie, dies of the illness, and also to information on George Orwell, who may have contracted the illness when living as a down and out and subsequently on to an e-book of his semi-autobiographical novel "*Down and Out in Paris and London*". Finally, you may be offered a link to the Public Health England website showing the inexorable rise in cases of tuberculosis in recent years.

Learning about mycobacterial infections in the past could have been an isolated experience. These days, it opens up the opportunity for interested students to explore fine art, theatre, fashion, baroque and contemporary music, opera and literature, as well as providing easy access to primary research data. Learning has become a rich experience full of an astonishing array of opportunities to explore the wider world that lies beyond the ivory towers of Academe.

If you are lucky, in addition to lecture notes and links to internet resources, you may get access to an illustrated set of notes, with images taken from the resources listed above. You may also have access, after the lecture, to a podcast to download so that you can listen again to the lecture; this is a particularly useful feature when planning your revision. Increasingly, lectures are being videoed and the videos are becoming available.

Periodically during your modules, you will be tested on what you have learned. Often, particularly during the early part of your course, tests will be in a multiple response format. You may be offered access to online question banks, similar to those used in the real tests, but which can be used as many times as you wish. After providing you with your score, the best question banks will also give access to explanations so that you can learn from your mistakes.

Numerous resources may also be provided to help you with your practical classes. In addition to electronic copies of the experimental protocols, there may be videos illustrating what you should be doing in your lab classes, animations illustrating concepts that are otherwise difficult to understand and links to software that will help you analyse and interpret your observations. Data generated by your classmates may be shared electronically. For experiments that are too complex or too dangerous for students to perform in person, simulations are available. In teaching infectious diseases, there are programs where clinical scenarios are presented. Make the right decisions and your patient survives: if you get it wrong, you could be summoned to a virtual inquest.

Virtual learning environments have more features. As well as sharing data, there are discussion areas, where students can post ideas and engage with discussion threads. Alternatively, collaborative work may be facilitated by "wikis" (named after the Hawaiian word *"wikiwiki"* meaning "very quickly"). These have the advantage that the editing history can be easily seen and the relative contributions of various group members may be assessed.

In contrast to journals, VLEs have tools for staff that permit monitoring of how, and when, resources are used and by whom. The scores of interactive tests available via the VLE are also stored, making the end-of-module assessment of student performance a relatively easy matter.

The resources described above are frequently referred to as contributing to **blended learning,** where face-to-face teaching is augmented by an array of online resources.

These undoubtedly enrich the student experience but they will probably never replace entirely teaching in lectures, tutorials and practical classes. The rest of this chapter is devoted to an exploration of traditional teaching methods and their associated resources.

4.2 Learning from lectures

The one-hour lecture still dominates the teaching of many courses; especially those taught to large numbers of students. If you think a lecture is the process whereby a lecturer's notes are transferred to a student notebook without the need for mental effort on anybody's part, think again. If you approach lectures in the proper manner, you will find that you will learn a great deal from them. The benefits that you can derive from lectures include:

- providing an overview or summary of an area of interest
- highlighting and discussing difficult areas
- covering and linking subject matter that is not well covered in standard textbooks
- introducing relevant research topics, which may include recent discoveries and other advances in your subject

4.2.1 *How do you get the most out of lectures?*

Lectures may be given in stuffy lecture theatres, possibly in semi-darkness. This, together with the time that they occupy, can lead to them making you feel very drowsy. Large lecture classes in particular limit the opportunities for discussion, making the flow of information seem to be only one-way, from lecturer to student. However, these will not be major drawbacks if you follow the advice below and learn how to *listen*. Listening in lectures should be an active process.

4.2.2 *Before the lecture*

To prepare for each lecture, you should:

- read about the general area of the lecture topic *beforehand*
- spend a few minutes before the lecture trying to recall what you already know about the subject – this will help greatly with a deep learning approach to your studies
- TURN YOUR MOBILE PHONE OFF BEFORE YOU ENTER A LECTURE THEATRE
- try not to arrive late. You will almost certainly miss a very important part of the lecture – its outline. Lecturers often spend the first few minutes of a lecture introducing the subject, and placing it in the context of other lectures. In any case, being late, unless for a good reason, is impolite
- sit near to the front of the lecture theatre if possible – you should be able to see and hear more easily
- do not chat, even in a whisper, during lectures. It is very rude, and it is highly distracting for others. Some lecturers may ask you to leave if you continue to talk during their lectures. More embarrassingly, they may want to join in your conversation...

- if you can, sit with a group of students with whom you feel at ease. If your lecturer breaks up the lecture into small group discussions, then this will be much more productive, and fun, if you get along with your neighbours

4.2.3 Active listening

Many lecturers use a standard format for their lectures; they will have a beginning (the outline of the lecture), a middle (the body of the lecture) and an end (a summary of the points that have been covered). Recognizing this form will enable you to organize your thoughts and structure your lecture notes.

A good lecture always starts with a **preamble,** such as … *"Hello, my name is Dr Frank N Stein",* and this preamble is followed in a good lecture by the **orientation.** This is where the lecturer tells you what will be covered in the lecture, and, possibly, how this material relates to other subjects given in the lecture course. Your notes can then reflect this structure as below:

Lecture on SUBJECT

    Orientation

    This lecture will consider the following key points:

        Topic A, Topic B, Topic C

        Topic A …

            Description …

                Extension 1 to Topic A

                First example (+ aside)

                Second example (+ aside)

        Topic B …

            Description …

And so on …until …

    The summary

Taking notes in a lecture is a very personal matter, but the following basic guidelines might help you structure the material more clearly.

Key points are the major topics around which the lecture is structured (Topics A, B and C). You should note key points carefully, along with their **extensions**. These are the secondary areas covered in the lecture.

Examples provide useful illustrations related to particular points in a lecture, and **asides** are less important points. These provide light relief in a lecture, and may be humorous in nature.

Summaries are used to bring together different sections of the lecture. These may often begin with a phrase like… *"what this goes to show is…",* or something like it.

Most lecturers provide opportunities for **questions** either during the lecture, or on an individual basis at the end. Don't be afraid to ask about things that you do not understand. Lecturers are there to help you to learn. Many students do not feel confident enough to ask a question in front of others during the lecture so, if you feel this way, you are not alone! However, if you are confused by part(s) of the lecture or want to know more, do go and meet the lecturer at the end and ask your questions then. Lecturers are usually very enthusiastic about their subject and welcome such student queries.

If you find that you have particular difficulties understanding an individual lecturer, because of the way the material is presented, it is a good idea to discuss these problems, if not with the lecturer concerned then with your personal tutor. Good lecturers will want to act on constructive criticism where there are positive suggestions about how to clarify the lecture material or improve the lecturing style. Do remember, however, that lecturers are human too and don't like negative criticism any more than you do, which is why constructive comments are far more likely to bring about the changes that you want.

Tips that can help you with active listening

Some lecturers provide a full set of notes, made available before their lectures; others may only provide brief outlines. Even if you do get given full notes, it is important that you listen actively to what is being said because it is extremely unlikely that a good lecturer will simply read out the notes verbatim. Annotate the notes during the lecture, taking down the "nice to know" material that may help you to remember the subject matter more easily and highlight key points. If you have access only to outlines, the following advice will help you to make the most of any lecture.

Be selective: you will not be able to take down everything in a lecture, unless you are very proficient at shorthand. If you are not selective in your note-taking, then you will miss important points, and may well end up misunderstanding the material being taught.

Take legible notes: it is more than likely that you will not have time to make "fair" copies of all of your notes, however much you promise yourself that you will. So, try to make legible notes from the outset. Don't forget to file any handouts with the relevant lecture notes after the lecture.

Reinforce your note taking: educational research has shown that when you leave a lecture theatre you will remember less than half of the content of the lecture. One week later, you will remember only half of this half. A few minutes spent in the evening of the lecture going over your notes can help to prevent the secondary loss, and this will save you much trouble later. Get into the habit of reading through your notes. To reinforce your reading, write an outline of the lecture, sketching out the main points, sub-headings and principal examples from the lecture. After you have done this, you should review your notes again, and amend your revision notes. List the points that you do not understand. If you find that textbooks do not help to clarify the problem, then ask your friends, or your personal tutor or the person who gave the lecture.

Link the material in each lecture: it is rare that lectures stand entirely alone. More often they occur within a module or course with common themes. It will greatly assist your deep learning if you search for and describe the links between various individual lectures, and then go on to develop links across the whole of a module or course.

Tips for making the most of your notes

Highlight essential points as you take notes, for example by <u>underlining</u> words or using CAPITALS or **asterisks**. If you carry a range of coloured pens, you may also wish to change the colour of your text in your notes. When you re-read your notes you can use coloured highlight pens, but don't overdo it since you can very easily overwhelm your text with fluorescent colours.

Compare notes with your friends. Everybody will bring a slightly different perspective to a lecture, and so it is likely that you will gain insights into the material that may have been important to your friends but that you considered of lesser importance. By sharing notes, you will expand your experience of the lecture material. You will also reinforce your learning through discussion with others, and this will be of great benefit when you want to revise.

If you make use of the advice above, you should enjoy your lectures, and you will be able to make the most of them.

Checklist for getting the most out of your lectures

- Arrive promptly or you might miss important introductory points.
- Do not try to write down everything the lecturer says.
- Listen to what the lecturer is saying and try to follow the thread.
- Make short notes to supplement the information in your handouts, when available.
- Note down any references to resources so you can look them up later.
- Do not chat your to neighbour – you will miss information and disturb others, including the person giving the lecture.
- As soon as possible after the lecture, look over your notes to remind yourself of the contents and look up any extra information you feel you need.
- Before the next lecture, make sure you have understood the information in the previous one. If you don't, you will find it more difficult to grasp new information.

4.3 Learning from tutorials

Tutorials offer a very special opportunity for learning, since they involve small groups of students working together. In this section, tutorials refer to groups of up to about ten students talking through a topic with a tutor. Typically, tutorial sessions last for about an hour.

Providing that the participants prepare properly for a tutorial, they can be amongst the most stimulating of learning sessions. At their best, tutorials promote discussion, clarify difficult topics and permit study of a subject at a greater depth than is possible in a lecture. They give you the opportunity to express *your* views and give your knowledge an airing. You can also practise your listening skills when others are making their contribution. Furthermore, they provide social occasions when you can get to know fellow students and your tutor better.

Successful tutorials require the **three Ps – Planning, Preparation** and **Participation.** You, the student, must take an equal share with your tutor in being responsible for the three Ps.

Planning involves finding out what is to be discussed well in advance of the tutorial session. You should discuss this both with your tutor and with fellow students. Tutorials are generally flexible and, with advanced planning, they can be adapted to reflect your needs and interests. Without planning, preparation becomes impossible.

Preparation for a given tutorial will vary depending on the work to be covered in that tutorial. It may require reading up on your lecture notes, or using textbooks and the internet to investigate a topic. You may need to interpret data that have been given to you before the session. Alternatively, it may require you to consider your own views on a given subject, and be ready to discuss them. It may just involve deciding the questions on which you particularly want to seek clarification.

Participation, by all of the students in the group, is the third pre-requisite for a successful tutorial. You must be prepared to make a personal input into tutorials otherwise sessions will either become a mini lecture or will develop into a dialogue between the tutor and the most vociferous member of the tutorial group.

Speaking up in a tutorial may seem difficult at first, but tutors do appreciate this problem, and you will have the support of your fellow tutees. After all, they are in the same boat, too. Never think that your opinions are worth less than those of anybody else. The best discussions build up in small steps, not as a series of brilliant flashes of inspiration. It is also a good idea to address your remarks to the whole group, rather than just to the tutor. In this way, everybody in the tutorial will become involved.

Remember

All participants share responsibility for the successful running of a tutorial. You, as a member of the tutorial group, must be prepared to play your part in making tutorials exciting and effective learning processes.

4.4 Learning from laboratory practical classes

Most science subjects are rooted in practical work, and much of our scientific knowledge has been built upon the deductions that scientists have made after analysis of experimental data. Therefore one of the primary objectives of practical classes is to develop a critical awareness of experimental method. Practicals have much more to offer, however. They teach: good laboratory practice, the necessity of paying attention to detail, and an awareness of using technical equipment appropriately. For example, microbiology students learn to handle microbial cultures safely using aseptic technique which, initially, is much more about protecting the cultures from their handlers than it is about protecting students from their cultures! Practicals are also used to demonstrate theories that you will have heard about in lectures. They may provide you with opportunities to exercise your problem-solving skills.

In many practical classes you will be working with a laboratory partner, and in some classes, you may need to form part of a team to tackle problems that are more complex. Acquisition of practical and team-working skills will benefit you enormously in your future career. Above all, safely run practicals are fun.

For each practical class you will almost certainly be given a handout or a workbook that describes the practical work that you will be undertaking and you should make it a golden rule that you **read and understand the introductory material that is provided BEFORE you arrive for a particular class**.

Highlight or underline any important points and make a note of anything you don't understand so that you can ask before you start the experiment. If you make this a habit, you will appreciate why things are done in a particular way and what is going on during the actual class. When recording measurements for representation as a chart, prepare a table and graph paper beforehand so you can easily and neatly note down your data as they are collected, and quickly plot them on a graph. This will enable you to identify anomalous points so that you can repeat them within the practical time allocation. **Note**: it is very good practice to keep a notebook specifically for laboratory work and to enter your data directly into it. For some types of class, books need to be retained in the laboratory. For example, when working with live cultures, it is necessary to minimize the risk of disseminating microorganisms. Maintaining a lab book helps you to learn the importance of recording your observations at the time they are made, and is excellent practice for when you are conducting independent investigations. By preparing for the practical in this manner, you will greatly add to your understanding and enjoyment of the practical session.

It may seem odd that we are advocating the use of old-fashioned lab books; why don't we suggest using your tablet device or smartphone? Undoubtedly these (or similar devices) will become the data collectors of the future but until they become cheap enough to be kept permanently in the lab, their universal use is probably a little way off.

The most challenging part of an experiment is obtaining a useful set of results. There are several steps to increase the likelihood that your experiment will be successful and these are outlined below.

> **Checklist for recording data**
>
> - Set any equipment appropriately and then make sure you use it correctly throughout the experiment.
> - Take accurate and careful readings.
> - When taking timed readings – take readings at the <u>exact</u> time required.
> - Pre-prepare a results table, preferably in a dedicated laboratory notebook, so that the data can be jotted straight into it. When appropriate, plot results on a graph as you go along – so that you can spot any anomalies and repeat measurements if necessary.

4.5 Field trips

Field trips allow you to make observations and to collect data in a manner that usually cannot be done in a laboratory. Because they are scientific investigations, all the information in the previous section is relevant; the laboratory notebook is, however, usually called the field notebook.

Results collected on a field trip will be influenced by a variety of external factors that you cannot control and are therefore likely to be more complex than those collected from laboratory-based practical experiments. Dealing with these uncertainties will give you more of an understanding of the issues that can affect scientific investigations. The effects of a host of external factors might require you to think about a topic in a wider context, encouraging you make connections between different areas of a subject. You will then gain a broader overview of the topic.

On a field trip you are likely to be working as part of a team with each group member making different observations. This means that you can assemble a much more comprehensive set of results than each person collecting results individually. Moreover, when teams work well together, the whole experience can be great fun.

Actively collecting data outdoors can be a much more memorable experience than collecting data in your usual practical laboratory so you are likely to remember more of what you actually did and the reasons why. Often, however, the most memorable parts of a field trip are the events that have nothing to do with your investigation, such as your best mate falling into a stream and getting soaked through, and these remain with you the longest!

Field trips are, by their very nature, organized outside the safe confines of a university building, and so all participants must be aware of potential hazards, how to avoid their dangers, and know what to do if things go wrong. So, in order to make a field trip a success, you should make sure that you are properly equipped with the recommended kit, follow any instructions you are given and ALWAYS do as your field trip supervisor(s) instructs. When assembling your kit remember to include:

- appropriate clothing, including long-sleeved shirts and long trousers which offer protection against the sun, stinging and prickly plants and biting and stinging animals. Remember, also, that several thinner layers are better than one or two thick ones.

- appropriate footwear; flip flops and high heels are unsuitable for walking far – Wellington boots are not designed for many country terrains
- relevant equipment, including pens and pencils, paper, field notebook, etc.
- communication equipment, although mobile phones may not work in remote places
- medication for any allergies that you may have
- sunscreen/insect repellent
- backpack and full water bottle

Do not be tempted to take prized possessions that could easily become damaged, such as an expensive camera or your favourite pair of trousers; their ruin could spoil your impression of the trip.

Other items such as specialized scientific equipment, first aid kits, compasses and survival gear should be provided by your university department or the field station.

Finally, remember that while you are on your trip you must take only the samples necessary for your work; you should cause minimal disruption that could harm any animals or plants and you must not leave any litter that could damage the environment. As the old saying goes: *"Leave only footprints: take only memories"*.

Checklist for a successful field trip

- Be aware of potential hazards and how to react in an emergency.
- Take appropriate items of clothing, equipment and medication.
- Comply with any departmental instructions.
- Follow instructions from your trip leader.
- Do not take prized or expensive personal items.
- Respect the environment you visit.

4.6 Teamwork

Much of this guide is about independent learning, but there are times when you will need to work as part of a team. At its simplest, teamwork involves learning to work effectively with a lab partner, as many practical exercises require that you work in pairs to complete the tasks on time. To be effective, both of you should understand the aim of the experiment and be prepared to contribute equally to the outcome of the practical. It is not unusual to see pairs of students conducting experiments in the lab, where one of the pair is relatively passive and doing very little, while the other is carrying out all the thinking required, as well as performing the bulk of the practical work. Although the passive member of the team appears superficially to benefit from this arrangement, the *most* benefit is derived by the active partner, who has learnt the skills of preparation, organization, time management, use of equipment, and so on. Maximum benefit will, of course, be obtained by both students if both play an active role in preparing for, and participating in, the practical session. The added bonus of following this advice is that you are also likely to find that your practical sessions will tend to go more smoothly and will be completed sooner.

4.6.1 Large teams

As well as working in pairs in practical classes, teamwork may also involve larger team exercises. Working as a group, you will typically achieve more than the sum of the individual efforts and so will be able to tackle bigger topics. By drawing on the different strengths of group members, individuals within the team can learn from one another.

All this sounds great but successful groups don't just happen. There are active steps that you can take to ensure the success of any group of which you are a part. Understanding the dynamics of the group is fundamental to this process and there is no magic formula that will work in every situation. This is because each group will comprise individuals each with different skills and personal strengths and weaknesses.

In every group, individual members will have individual responsibilities. Each person will have:

- specific tasks to accomplish on their own which need to be brought back to the group
- a contribution to make to the work done by other members of the group; most often, this is done by *constructive* criticism of the efforts of other team members
- a responsibility to ensure that the team works effectively together

And, within the team, there are three interrelated areas of activity:

- task-related activities, directed to achieving the objective set for the group
- group-related activities, directed to ensuring that the group works together effectively
- individual tasks for group members

Effective groups allow individuals the freedom to achieve their own objectives, thus allowing the group to achieve the overall objective. In the most effective teams, all members are aware of their individual responsibilities, the responsibilities of other members, and the responsibility they owe to the group. Any individual may take on more than one role.

To be effective, every group needs:

- a team leader, or at least someone to co-ordinate the activities of the group. This person is responsible for allocation of tasks and for ensuring that good communications are maintained with all team members. If the final product of the team exercise is written communication, the team leader might, for example, take on the role of editor to ensure a uniformity of written style is imposed.

- good communication with all team members. Every member of the team should be present at team meetings, so these need to be arranged for times when everyone can be present and at a venue that is accessible to all members of the group. Any job inevitably takes longer than envisaged at the outset and deadlines have a nasty habit of creeping up rapidly. So, to avoid unpleasant shocks, don't leave everything to the last minute.
- shared responsibility between all team members for all of the output from the team effort. Just because you are not primarily responsible for a task, this does not mean that you should not help to improve it if it has not been done well, and don't just blame the individual whose primary responsibility it was. Ultimately, who is going to look bad for inferior work produced by your team?
- allocation of tasks according to the specialist talents of *its various members,* although it is important to be flexible in the roles that you take on when engaged in teamwork
- good group dynamics, where team members are supportive towards the others on the team, are unselfish, and do not form cliques within the group

Consider the following example: a team has three members who make an average effort and is led by a member who is outstandingly productive. Despite this, however, there is one member of the group who never answers emails, does not turn up for group meetings and has not produced a single piece of work. Clearly it would be unfair for that individual to benefit from the hard work of the others. It is particularly hard on the group leader who is otherwise dynamic and effective.

Sadly, instances like this do occur. Where appropriate assessment is in place, group members would be given the opportunity to penalize team members who do not contribute fully to the work.

Checklist for effective teamwork

- Good communication – this underpins every successful team.
- Understanding group dynamics – this will help you to make the most of any teamwork in which you engage.
- Recognition and utilization of the different strengths of individual members of the team.
- Team members who fully engage with the task and take their responsibilities seriously.

4.7 Books

4.7.1 Buying books

If your personal study is to be effective, you will require a number of textbooks. School courses tend to work from defined set books but for most university courses you will have greater freedom of choice. Which books should you buy, should you then pick an e-book or a paper copy and from where should you buy them?

Choosing a book

You will probably be recommended to consider buying a variety of books. One of the first steps that you can take to help you make the decision on what to buy is to talk to last year's students about different texts. Was the coverage insufficient, or was it over-the-top? Was it up-to-date? Was the course designed with that book in mind, or another book, or none at all? Talk to the people who teach on your course, and check that the book you propose to buy is suitable. It is also important to check that you will not be required to work from a particular set text if you decide that you do not want to buy it.

You should also visit the library and spend a little time looking at the array of books that they have on related subjects. Scan through them to get a "feel" for the book. You need not attempt to understand the whole content of these books at first. Do you find the text style readable? Are the illustrations clear? Is the text properly sub-divided into sections with appropriate sub-headings, or are you faced with a monotonous sea of print? Does each section have an introduction and a summary of the material it contains? Dip into random paragraphs to see if they are too long. What do you feel about sentence length? It can be very difficult to read long, complex sentences that contain several ideas, but when people are writing texts they may get so caught up in their ideas, which are, of course, so fascinating to them that they forget the poor reader! See what I mean? Ask yourself: "...*would I enjoy working from this book*?"

When it comes to the question of 'real' books made from paper or e-books, the choice is largely one of personal preference. If you opt for an e-book, you will be constrained to using a particular reader, although this will store all of your textbooks in a single location. You will also not be able to sell your book on once you have finished with it. Physical books take up a lot of space and can be very heavy when you move. However, you can easily lend your copy to your mates.

Actually purchasing your books

You can either buy new or second-hand books. Many university towns and cities are fortunate in having a number of bookshops, but there is little point in shopping around as the price of books hardly varies between them. Alternatively, it is sometimes possible to purchase books from internet sites at discounted prices. At this point, however, a word of caution is necessary. Some bookshops stock expensive hardback books when cheaper paperback versions are available. The difference in price between a paperback and a hardback book can be significant and so you may end up paying quite a lot extra for a couple of bits of board! It is also a good idea to check that a new edition of the book that you want is not just about to be published.

Students in higher years frequently sell textbooks that they used earlier in their studies, often through advertisements placed on notice boards. Buying books in this way can save considerably on the cost of a brand new textbook. Before buying, however, remember to use the guidelines above to judge whether the text would really suit your purposes. Ask yourself why the book is being sold. You should expect second-hand textbooks to show signs of wear and tear, but if it is sold "as new", was it *really* useful for the course for which it was recommended? Don't buy a second-hand book solely on the recommendation of the seller.

Before you buy

Check the edition of the book being sold. Some subjects like molecular biology advance at an alarming rate, and at higher levels, a three-year-old text may be hopelessly out of date. By contrast, anatomy books, for example, would not be expected to change at the same rate. If the textbook on offer does not appear to be the latest edition, then be very cautious about buying it. If you have any queries about buying a book, you should consult the person running the course.

4.8 Using libraries

Libraries contain tens of thousands of books and journals, and their very size can be daunting when you first visit them. Even students who are used to libraries at school or in their home towns can feel intimidated. Fear not, help is at hand. University libraries are organized to assist you with your studies, and there will be a range of resources to help you with searching, evaluating, managing and referencing information, as well as advising how to keep up to date with new information. They will also be able to advise on copyright and plagiarism issues.

4.8.1 *Making the most of the library facilities*

There are three main reasons why you will use the library: quiet study, consulting or borrowing textbooks, and looking up reference material in journals. Within the libraries are desks and tables that help you study in peace and quiet.

Broadly speaking, there are three types of reading material to be found in academic libraries; textbooks, advanced books on specific academic topics, and journals. In general, textbooks are collected together in different sections of the library by subject and "classmark", and are stored separately from journals.

Many universities make a point of introducing new students to the relevant libraries, showing them where material is kept, how to access it and the various rules of the library. This introductory visit will certainly cover how to find what you are looking for – so pay attention since this advice will prove invaluable later on in your studies!

Borrowing books – and fines!

Most textbooks and sometimes bound volumes of journals may be borrowed from the university libraries. Most books can be borrowed for extended periods but there are some that are in high demand and may only be borrowed for a limited time to allow others fair access to the material. Books in really high demand are loaned out for just a few hours. This is usually because a book is required for a particular course and so, for a limited period, a large number of students will need access to it. In these situations, the book is transferred to the "counter collection", which means that it is only available on request from the librarian on the counter. This allows all students the opportunity to refer to the book, rather than the lucky individual who happens to borrow it first.

Overdue books and journals attract fines that can quickly build up to significant and alarming sums of money. As well as being potentially expensive, it is bad manners to keep a book longer than necessary, since others may require access to the same text.

If a library book is not available when you first want it, it is sensible to ask for it to be reserved. As well as ensuring that you eventually have access to the book, it also alerts library staff to the demands for that book. This may mean that the library buys additional copies for future use.

Ask for help

A library can be a daunting place when you first start to use it. When starting out, use the library catalogue (it will be available online) when searching for specific books and journals.

Will you make mistakes?

Will you fail to find references?

Will you need advice on the best way of tackling a particular search?

Yes, Yes, and Yes. Do not despair. The library staff are highly trained and are expert in such matters. They are there to help you, so don't waste hours getting nowhere – ask them for help! They will be pleased to point you in the right direction, and to show you where you have gone wrong. This will help you to avoid those pitfalls next time you need to search for information.

4.8.2 Journal papers and how to understand them

There was once a pathologist who would come into his laboratory every Friday morning, hand over the latest copy of a famous medical journal and say to his staff: "Well, here are this week's fairy stories". Although the pathologist was making a joke there is an element of truth about what he said.

At school or college, textbooks will provide you with all the information that you need to excel in your subject. At university, while you may get an excellent grounding in the subject from textbooks in your first year, you will soon need to start reading, and understanding, scientific papers.

Scientific papers are written in a highly structured, conventional manner. It is important that you appreciate how the conventions work if you are to get the most from the papers, so we need to start by considering how papers are written and published.

A journal is a regular publication (perhaps weekly or monthly) in which academics and researchers will publish their research findings. Each published research study is called a "paper" (in science, it is called a "scientific paper") and is extremely detailed so that the reader can repeat the work if necessary. We will talk more about research papers below. These days, journals are nearly always also available in electronic format and so academics and students can read journals from their computer rather than having to visit the library to do so.

University libraries subscribe to a large number of journals devoted to a staggering array of subjects. Current issues of the journals (i.e. recent issues produced that week or month) will be found on the "current journals" shelves of the library until a volume (usually a complete year's worth of individual issues) is completed. Shortly after the volume has been completed, it will be sent away to be bound as a hardback volume. The "bound journal" is then returned to the library, and is stored alongside earlier volumes of the same journal. The various bound journals are arranged alphabetically by journal title and are also placed in date order as you would expect.

As well as journals being listed in the relevant library catalogue, at the end of each bay are notices saying which journals it contains (e.g. *Nature or The Lancet*). It goes without saying that electronic journals don't disappear for binding!

4.8.3 How scientific papers are published

Scientific researchers aim to publish the results of their work in high quality journals with a large readership. The work is written up in the form of a "paper" and then submitted for consideration by their preferred journal. These papers are usually sent to two or more referees for "peer review" before they are published. Peer review means that the submitted paper will be reviewed by several other academic experts ("peers") who will be able to comment on the quality of the paper and spot any issues that need to be addressed. Their job is to take a critical view of the material that has been submitted for publication and then to write a report to the editor of the journal, pointing out any matters of concern.

The sorts of questions asked during the peer review process include:

- are the research findings of the author(s) really original?
- do the experimental results described fully support the conclusions drawn by the author(s)?
- is the paper clearly written so as to be unambiguous?
- is the paper comprehensive so that the experiments can be repeated by the reader to test their validity?

...and so on...

Only if a paper meets these criteria will it be published in a peer-reviewed journal.

This thorough approach generally works well, but the system is not perfect and mistakes do creep into the literature.

There are three major types of paper: full original research papers (often collectively called the "primary literature"), short "research letters", and review articles. These are now considered in turn.

Primary literature

Most of the papers that you will come across are full descriptions of original experiments made directly by researchers. Although the details differ between journals, a typical scientific paper includes:

- abstract – a very brief summary of the work, the major results, and the author's conclusions
- introduction – a short overview of that specific field of research and where this new research fits in
- materials and methods – a concise description of the methods used to undertake the research
- results – a detailed presentation of the original experimental results; depending on the experiments, this section may also include statistical analysis (see *Catch Up Maths & Stats* for details on why authors use particular statistical tests to analyse their data)
- discussion – a description as to what the author concludes that these data show
- references – a list of the other research papers that the author refers to in his work (allowing a reader to look further into particular aspects of the topic)

Research letters

When results, sometimes only from single experiments, are too important to wait for the whole body of work to be completed before publication, then these smaller sets of results may be released as research letters.

Alternatively, a research letter may be a critique of work that has already been published, perhaps including data to refute the previous paper. In such cases, the author(s) of the original work are often invited to publish a reply, most often appearing next to the first research letter.

Review articles

Given the very high volume of research papers published, it would be a huge task for a student to read all the relevant papers in a subject area. This is when review articles can be helpful, particularly when finding your way around an unfamiliar area. Therefore, although we have described reviews last in the discussion of the three types of journal, they should often be the first place you start when learning about a new subject area.

Reviews are typically written by experts in their field and give either an in-depth view of a particular topic or present a broad overview of their field of interest. Reviews cite many primary publications (papers describing original research) in support of their arguments, and interested readers can use these to explore the literature further, often with the benefit of critical comments made in the review.

A variant of the review article is the short "news and views" type article, often devoted to a critical review of a specific paper published in the same edition of the journal. "News and views" articles tend to be commissioned by the journal from leading experts in the field.

4.8.4 *How should you read a scientific paper?*

How you read a scientific paper (i.e. a paper from the primary literature – see above) depends on what you want to do with the information. Early in your undergraduate career, you should aim to get a feel for what is being published in an area. However, by the time you undertake a research project in your final year, particularly if it involves experimental work, you will need to pay much more attention to the way the science was done.

A good approach to take when faced with reading a scientific paper is to go through it in the order of the steps listed below.

Step 1

Look at the title which is a crucial part of any paper. As many papers may be published on a topic, authors often try to make their titles interesting, to encourage the reader to explore further: "*GM foods – a case of resistance*" is an allusion both to the antibiotic resistance markers left in first-generation insect-resistant plants and to the public opposition in the UK to the introduction of GM plants into commercial agriculture. If you come to write papers, you will need to remember that the title may be the only part of the paper that many people will read.

Step 2

Read the abstract or summary. This is a short description of the principal findings of the paper, and, together with the title, gives the reader a feel for the subject. For many, reading the abstract will be sufficient, but if this is all you read you may miss out on important details that lie hidden in the body of the text. It is not always wise to rely absolutely on the authors of a paper to highlight in the abstract what is important and "hidden gems" may be found by those who are prepared to explore deeper into papers.

Step 3

What you read next depends on why you are reading the paper and what you want to do with the information. For someone who is relatively unfamiliar with the research in that area it may be invaluable to read the introduction section next since this sets out the background against which the research was carried out and introduces the topic in context. People already familiar with similar work may not find the introduction of much interest and might skip this step.

Step 4

The next step is to read the results that are reported in the paper and the interpretation placed on them by the authors in the discussion section. Typically, "Results" and "Discussion" sections are written separately, but when complex sets of experiments are being reported, particularly if the implications of one set of results need to be explained before the next set can be understood, a combined "Results and discussion" section may be provided.

Step 5

The papers to which the author(s) refer are described in the reference list and you might want to follow up the information in the paper by reading some of these, especially if you are preparing a dissertation or literature review. Finally, the acknowledgements section of the paper is typically very short and often need not be read by the students. However, it contains important information, particularly concerning the way that the work was funded.

Step 6

If you wish to do similar work to that described in the paper, you will need to read the "Materials and methods" section. In places, the "Materials and methods" section may not describe a particular method used in any detail, but instead refer to other papers in which this is described.

4.9 Using the internet as a source of information

Although it now seems almost inconceivable, not many years ago, the only way to follow scientific progress was by reading the published journal papers in libraries. The advent of powerful and widely accessible computing and the existence of electronic journals has changed all that, for the better. Now, locating the journal articles you need to read and then reading them is fast and easy, if you know how!

When you need to search the primary scientific literature, a number of online tools can help. It is important that you learn to use them as they will undoubtedly make your information searches quicker and easier. Three particularly helpful tools are:

- Web of Science (http://isiwebofknowledge.com/) – the Web of Knowledge covers many thousand journals but allows users just to access the relevant scientific articles via Web of Science. Coverage does, however, get patchy, the further back in time you need to search

- PubMed (www.ncbi.nlm.nih.gov/pubmed/) – this is a service of the US National Library of Medicine that includes millions of citations from Medline and other life science journals for biomedical articles back to 1948. PubMed includes links to full text articles and other related resources

- Google Scholar (http://scholar.google.co.uk/) – provides users with the opportunity to use a range of search options across the scientific and medical literature

These three database "search engines" allow users to search for work published in journals using key words, authors, journal title, etc. In addition, most search engines like these now allow you to click on links in the reference sections of papers and you will be taken to the full text of that article too, allowing you to build a web around the original paper by following links forwards and backwards.

None of these tools is difficult to use, and they can prove invaluable, particularly during project work. When you first use these databases, it is easy to get overwhelmed by the number of citations that you find. All of these databases do, however, allow you to refine your search, narrowing down the list of papers, perhaps to just the one that you need. You may do this by combining different keywords, author names, journals, year of publication or any combination of these, providing, of course, that you have this information to hand.

Coverage of the literature is constantly being updated, so you will find new papers almost as soon as they are published. Indeed, with some journals, you may have access to the electronic version of an article before its paper version appears from the printing presses. The coverage of older papers is variable, depending on the database, but for most there is now reasonable coverage of journals published from the 1960s to the present day. Many of these databases are undergoing frequent redesigns so we have not provided detailed advice on their use – it is all fairly intuitive.

Assuming that your university library subscribes to the journal containing the article found by your search, then you will be able to have free access, from your computer, to the full article, so there is no need to go hunting through the library to find the paper copy. If your library does not subscribe to the journal (although most university libraries do subscribe to a huge number, there will be some to which it does not), then you will still be shown the title and abstract and offered the chance to buy an electronic copy. Before you buy, check the abstract to see if it really is a paper you need and then talk to the librarian to see if there is a cheaper way to gain access.

There are a number of other similar tools, which you are free to explore and your librarians will give you advice on how to get the most from them.

Remember

Databases are only as good as the data that they contain and they don't all cover all of the journals. It is advisable to use more than one resource when researching topics.

4.9.1 *Finding reliable internet resources*

This brings up the whole question of how reliable the internet is as a source of information. It is **essential** to remember that it is very easy to post information on the web and many people do. The reasons people put up information are many and various; not all of them are for the greater good. Even apparently reputable websites may be designed to give a particular "spin" to a topic.

You need to be very very careful when evaluating information from the internet, particularly about the **structure and content** of the material and also what **authority** the website has. Except for web pages in academic journals (which includes papers identified by the Web of Science, Medline, PubMed, etc.) other web pages are rarely "peer-reviewed". That means that it is down to you, the student, to evaluate which web pages to trust, and which to reject. The questions asked during the process of peer review (see above) may help you to make the right judgements. Here are some further questions to help decide:

- is the information relevant and pitched at the right level?
- are fact and fiction clearly differentiated?

- are references provided?
- are there links to independent sites in support of the content of the site in question?
- has the information been validated in any way?
- is the site up-to-date?

Sites that are not supported by valid references or external links or those that are not substantiated by peer review are more likely to be inaccurate and/or biased.

When it comes to considering the authority of a website, you may also like to bear the following in mind. A website produced by a corporation (with a web address ending in ".co" or ".com") may give a view that favours that company. After all, companies exist to sell their products. In contrast, university sites in the UK contain ".ac.uk" and in the USA they end in ".edu". UK government sites contain ".gov" or ".org".

However, just because a website is not a company site does not guarantee reliability of the information. You should consider:

- who wrote the material?
- why was it written?
- are the credentials/qualifications of the author shown and relevant?
- is the site associated with a reputable institution (universities/research institutes/professional associations/government bodies)?
- can you see contact details for correspondence?

Wikipedia

In their early years at university, many students use Wikipedia as a source of information. This dynamic online encyclopaedia that anyone can edit contains many good pages which have external references and these are a useful resource. It claims to have a neutral viewpoint, its content is free and it has a code of conduct, part of which states that Wikipedia has no firm rules. You should, however, be **very careful** how you use Wikipedia. Although it is a good secondary source of information and as such will provide background information on a huge array of topics, it should not be quoted as a primary reference because it is not peer-reviewed and anyone can edit its pages, leaving open the possibility of unreliable information being posted on it.

Looking at the "discussion" tag may give some idea of how reliable the information is on any given page, particularly if it concerns a controversial topic. Furthermore, information may change after you have read a page so any reference you quote may change and others will not be able to access the information you cited. This means you need to take care to provide a **full** reference to the **actual page** you accessed. These may be traced using the "history" tag. Here is an example of how to do this:

WIKIPEDIA. 2014. *Charles Darwin* [online]. [Accessed 20:30 hrs Monday 12th May 2014]. Available from http://en.wikipedia.org/w/index.php?title=Charles_Darwin&oldid=608251887. 2014. The link to the current, live page is simply: http://en.wikipedia.org/wiki/Charles_Darwin.

Even if you can satisfy yourself that the website looks fine, remember that, just occasionally, science undergoes seismic changes.

In 1978 Peter Mitchell was awarded the Nobel Prize in Chemistry for expounding the chemiosmotic hypothesis to explain how mitochondria generate ATP. A few years previously he was not at all well regarded by some scientists in his field. Suppose he had had a website with the same content before and after he won his prize. How do you think that these would differ? How do you think your views would have changed on what you read?

Chapter 5 Presenting your work

5.1 Plagiarism

Material obtained from textbooks and scientific papers will, in many cases, form the basis of a considerable part of your work. You **must** acknowledge the work of the original authors with a citation and a reference, so that your reader knows the source of your information and can check it. Plagiarism is defined as presenting someone else's work as your own. Here, work means any intellectual output and typically includes text, data and images. It is likely that you will be asked to provide an electronic copy of an assignment so that it can be assessed for originality by software designed to detect plagiarism.

Here are two examples of using a source text: one illustrating plagiarism; the other showing creative thought. These are presented together with the extract that was used as the source. Version 1 is from page 4 of an article: "*The Future of Genetically Modified Crops: Lessons from the Green Revolution*", by Felicia Wu and William P. Butz, also at http://www.rand.org/content/dam/rand/pubs/monographs/2004/ RAND_MG161.pdf [Accessed 12/5/14].

Version 1 – the original paper

"It is in the context of the hundred-year history of technological change that we consider the most recent movement in world agriculture: genetically modified (GM) crops, produced through modern biotechnology that enables genes to be transferred across different species and even across different plant kingdoms, to introduce desired traits into a host plant. After just a decade, the GM crop movement is already beginning to revolutionize agriculture in new ways, with previously unachievable benefits and novel potential risks.

The study of GM crops focuses on the genetically modified crop movement and whether it has the potential to revolutionize agriculture in the developing world and to truly become the "Gene Revolution" that some of its proponents already call it. We focus on the developing world because it is in greatest need of a new agricultural revolution – whether in the form of GM crops or another revolution altogether – given the rapidly growing populations, lagging agricultural technologies and malnutrition in the world's poorest nations.

Three presumptions motivated this study: (1) reducing hunger and malnutrition is desirable; (2) now, as in the past, revolutionary technological change in world agriculture can substantially reduce hunger and malnutrition; and (3) now, as in the past, agricultural technologies can be designed and used such that the majority of farmers, consumers, and experts will agree that the technologies are worth their attendant risks."

Version 2

"This report considers the most recent movement in world agriculture within the context of the hundred-year history of technological change. Genetically modified (GM) crops, produced through modern biotechnology, enable genes to be transferred between different species and even different plant kingdoms, so that desired traits into a host plant are introduced. It has only taken little more than a decade for the GM crop movement to revolutionize agriculture in new ways, and although there are now many previously unachievable benefits, these have been accompanied by novel potential risks. The report will focus on whether these new technologies have the potential to revolutionize agriculture, particularly in the developing world, thereby becoming the "Gene Revolution" in contrast to the "Green Revolution" that took place in the mid-twentieth century when the introduction of intensive use of agrochemicals, selective breeding and improved agricultural management, boosted production. There will be particular focus on the developing world – the world's poorest nations – because it has greatest need of a new agricultural revolution to try to avoid disaster for the rapidly growing populations which are under threat of malnutrition due to backward agricultural technologies."

Version 3

"Since the early 1900s, an increasing number of innovations and developments in agricultural technologies have led to massive increases in crop production. The introduction of intensive use of agrochemicals, selective breeding and improved agricultural management, in the mid-1900s, boosted production and created what is now dubbed the "Green Revolution". The next great advance is now coming through improvements in the ability to manipulate the genetic material directly, and during the past ten years these scientific techniques have opened the way for the successful transfer of specific desired phenotypic traits not only between many different species, but also extending between different plant kingdoms to produce genetically modified (GM) crops. This progress has been likened to a "Gene Revolution"! However, any perceived benefits of GM crops are accompanied by real potential risks, and the cost–benefit debate must take place to ensure that these new technologies do not create further problems for the future.

Parts of the world that are in most need of agricultural improvements are centred in the poorest nations, where, typically, underdeveloped agricultural technology is unable to cope with the increasing population, resulting in widespread hunger and malnutrition. This report will, therefore, focus on the potential benefits that the new GM crops might bring to these Third World regions and investigate the methods by which any accompanying risks can be limited."

Version 1 is the original text. Version 2 shows several changes, mainly with different words being inserted but with some original additions. It is not sufficiently different from the original to avoid the accusation of plagiarism. In only one place, when the "Green Revolution" is mentioned, does it show any original thought. Thus, it is considered to be plagiarized from the first version, even if a reference to the original text had been included. Version 3 shows independent thought, does not use the same sentences or phrases, except when describing common terms (the "Green Revolution", the "Third World" etc.), presents several points in a different order and makes connections to other ideas to explain the context of the topic. This would still need to include a reference to the original paper since this is where some of the ideas originated (but were not copied verbatim).

AVOID PLAGIARISM

It is never a good idea to copy out large sections from a textbook or other source. Apart from the fact that this is plagiarism, you will derive very little educational benefit from such an exercise. You may also get caught, and you will be heavily penalized. In the most extreme cases you may be expelled from your university. As work is routinely submitted to a plagiarism detection service, there is a very high chance that, should you be tempted to pass off the work of others as your own, you will be caught. It is far better to show how you can relate ideas to one another. At university, it is this higher level of understanding that we are trying to develop.

5.2 Good writing and writing style

The ability to present information effectively is a skill that will be useful throughout your life, and efficient communication is essential if you work within a team. At university you will be required to write essays, practical reports and projects, to give presentations as talks and possibly as posters. Many of these require common skills, albeit tailored to the specific task. These include planning, preparation, organization and an understanding of your audience. The way that you present not only reflects what you know about a subject, it can also show how well you can analyse a topic and integrate it with other areas of thought or knowledge. A recurring theme present in all types of good presentation is the organization of the material, where material is grouped into a beginning, a middle, and an end.

Lewis Carroll summed up the process in *"Alice's Adventures in Wonderland"*:

> *The White Rabbit put on his spectacles. "Where shall I begin, please your Majesty?" he asked.*
> *"Begin at the beginning," the King said, very gravely, "and go on till you come to the end: then stop."*

The manner in which material appears on the page, or screen, can also significantly affect how it is received, and you should consider this whenever planning written work for assessment so that you present your work in the most appropriate and interesting way.

5.2.1 Third person, passive voice and the past tense

Whatever you have been told at school or college, your writing at university level will sound more authoritative and objective when written in the "third person".

Instead of writing "*In this essay I will describe...*" it is better to write "*This essay describes...*". This is particularly important if you have been invited to write critically about a subject since you need to draw unbiased conclusions in such essays.

Many writers, especially those in scientific disciplines, write in the **passive voice**, as in the following example: "These organisms were classified by Margulis". However, excessive use of the passive voice leads to reader fatigue and a dull and lifeless text. To make it read better, a mix of passive and active voices should be used, as in the example: "After Margulis had completed his classification of organisms, the research was continued by Schmidt".

Write practical/laboratory reports in the past tense. This is particularly important in sections on methodology, as if you use the present tense, e.g. "the grollifying cordwanglers are added at this point", it sounds like a recipe.

5.2.2 *Sentences*

Verbs and nouns

Every complete sentence MUST have a verb – an 'action' word. The simplest sentence contains a single word of command. An example would be "Go!". Generally, however, sentences comprise at least a noun and a verb. "John laughed" or "Pigs fly" are examples of very short sentences. The verbs, "laughed" and "fly", are action words. In these cases, the noun, "John", is a proper noun – it is the name of something (John) – and so its first letter is written using an upper-case character regardless of where it comes in the sentence. Ordinary nouns, such as "pigs" don't get such special treatment unless they start a sentence.

Subjects and objects

In the sentences above, "John" and "pigs" are also the "subjects". Occasionally, the verb is referred to as the predicate of the subject, since it tells you something about the subject. Most sentences also have an object; this completes the meaning of the verb. The "object" in the sentence "John laughed at the joke" is "joke" and in "Pigs fly home" it is "home".

Pronouns

Sometimes pronouns are used in place of nouns, for example, I, you, he, she, it, we, you, they, yours, mine, this, that, some, none, and so on. "John laughed at the joke" could be written "He laughed at it", where "he" and "it" are pronouns. Pronouns change, depending on the context in which they are used. When it is the subject of the sentence, "I" is the correct personal pronoun but when "I" becomes the object of the sentence, it changes to "me", for example, "I laughed at the joke", but "John laughed at me".

One area where this causes major confusion is in the use of "who" and "whom". In the sentence "Who is there?", "who" is the subject; "with whom you have spoken" is a phrase where "whom" is used, correctly, as the object of the sentence. Nevertheless, it sounds a bit contrived and it is doubtful whether these days you would hear the latter phrase often spoken. The grammar checking facility in *Microsoft Word* often confuses the use of "who" and "whom".

Adjectives and adverbs

Adjectives describe nouns; adverbs do the same for verbs and together they make sentences far more interesting and explanatory. In the sentence "Aerodynamic pigs fly acrobatically home" there is one adjective "aerodynamic" and one adverb "acrobatically", whereas the sentence "Jovial John laughed merrily at the funny joke" contains two adjectives, "jovial" and "funny" and one adverb, "merrily".

Prepositions

Prepositions are words that link nouns, pronouns and phrases to other parts of a sentence. These are the small words, like "in", "at", "under", "on" and so on. The preposition "in" is used in the following sentence "Jovial John laughed merrily at the funny joke in the book" and for the sentence "Aerodynamic pigs fly acrobatically home at bath time", the preposition is "at".

Some grammar 'experts' frown on the idea of ending a sentence with a preposition. Trying to adapt sentences to avoid doing so can sometimes lead to nonsense, as Winston Churchill is said to have remarked to an editor who had clumsily rearranged his prose, "*This is the sort of English up with which I will not put*"[1]. Common usage often means that your sentences do not read too badly if you do end them with a preposition, but it is usually possible to rearrange your sentence so that everything is grammatically correct. Using Winston Churchill's example you could write "I will not put up with this sort of English".

Conjunctions and interjections

Conjunctions join different parts of sentences together; "and" and "but" are good examples, as are "or", "nor", "if", "when", "whereas" and "because". Our two example sentences actually consist of two separate ideas which can be joined by a "because" and "when" respectively:

"Jovial John laughed merrily *because* there was a funny joke in the book"
"Aerodynamic pigs fly acrobatically home *when* it is bath time"

Interjections such as 'Ah!' or 'Oh dear!' are added to a sentence to heighten the dramatic impact and are often accompanied by exclamation marks! You can use these too liberally! Then your reader may end up feeling as though you are shouting! They should generally be restricted to the spoken word.

Split infinitives

Returning to a consideration of verbs, some people with a classical education will tell you that it is "wrong to split an infinitive". The infinitive form of any verb is "to [the verb]". The most famous split infinitive is "...*to boldly go*...", where the adverb "boldly" splits the infinitive form of the verb "to go". Although this construction is

---

1 Sir Ernest Gowers' *Plain Words* (1948)

not actually wrong, a sufficient number of people feel that is a very inelegant way of expressing oneself. For that reason alone, it is best avoided whenever possible.

5.2.3 Clarity

This means using plain, simple words where plain simple words will do. Simple sentences and simple paragraphs also help with clarity of style.

In general, it is preferable to use short sentences but, if you write longer ones, join phrases together with conjunctions; semicolons are also useful. The previous sentence is an example of how long sentences can be compiled. Paragraphs should be given over to one idea, and have an introductory sentence and a concluding remark.

Read the two paragraphs below. They illustrate what clarity *ought* to be about, even if they only show what does **NOT** constitute good style.

> *Simple sentences contain only one idea. They avoid multiple subordinate phrases. Simplicity can be overdone. Constant use of simple sentences tires the reader. Use of too many simple sentences introduces ideas too rapidly.*

> *Alternatively, if authors want to impress their audience, particularly if they have literary pretensions, they may exhibit a desire to show off their expertise at constructing complex word orders, with the effect that by the time the reader has finally and terminally managed to sort out the point of the sentence, then they have forgotten what the beginning of the proclamation was about, and will need to re-read the beginning of the sentence to remind themselves.*

You should avoid paragraphs that comprise a single sentence!

To achieve clarity of style, you should try to avoid:

- unnecessary phrases that add words to your argument but not content, such as "in addition to" where "also" will do
- tautology, defined in the *Oxford English Dictionary* as: "the repetition (esp. in the immediate context) of the same word or phrase, or of the same idea or statement in other words: usually as a fault of style". Examples include "to encode for" or "reiterate again". The first may be expressed as "to encode" or as "to code for". In the second example, the word "again" is unnecessary
- inclusion of words such as "...clearly..." or "...obviously...". What may be clear or obvious to you may not be clear or obvious to your readers; the last thing you want to do is make your readers feel stupid

5.2.4 Accepted usage and jargon

Accepted usage means that grammatical constructions that were unacceptable in the past become so prevalent that (almost) everyone now wonders what all the fuss was about. Sentences that end in a preposition ('about'), like the last one, are a good example. A grammatically correct form, acceptable to purists, would be "...become so prevalent that, about what, everyone wonders, was all the fuss"!

Laboratory jargon also sometimes seeps into English usage. A fine example of seepage is the term "sequencing". The *Oxford English Dictionary* definition is "to arrange [objects] in a definite sequence or order". In relation to nucleic acids, however, (almost) everyone now talks of "sequencing" when talking about discovering the order in which nucleotides are already arranged.

5.2.5 Over-use of adjectives

This can also lead to a lack of clarity when they are piled up before the noun to which they refer. Sometimes these have the force of accepted use, such as the bacteriological term *"extended-spectrum β-lactamases"*. However, on other occasions this practice can be misleading. Does *"muscle wasting treatment"* mean a treatment that causes or cures muscle wasting? Appropriate use of a hyphen can help: *"muscle-wasting treatment"* but why not be unambiguous and, instead, state *"treatments for muscle wasting"*?

> **Remember**
>
> Clarity and precision of communication aid critical thinking about the subjects being communicated, so the words you use must communicate what you mean to convey.

5.2.6 Legibility and quality of presentation

These are undoubtedly of considerable importance in helping students obtain high marks. Coursework these days is usually produced electronically, but legibility is essential for essay-style examinations. Legibility is important not because markers deliberately deduct marks for poor presentation or legibility, but because it is much easier to follow an argument if you are not continually being distracted by attempts to translate the hieroglyphics. When producing coursework, make sure that you always use *at least* 1.5 line-spacing. Use double spacing if instructed so to do. This permits written comments to be added to your work.

5.2.7 Waffle and irrelevant material

Waffle is the repetition of a topic over and over again in slightly different ways, without any new information being presented, thereby destroying the continuity of your argument. It is head of the hate list for all markers faced with a pile of essays or practical reports and should be avoided at all costs. It is much better to be concise, but try not to be too blunt.

Inclusion of irrelevant material is a common fault and occurs when students write about a topic they *would have liked* to write about and that was *nearly* the same as the topic they were actually set. If, for example, you are asked to write an essay on frogs, you will not get any marks for writing an essay about toads just because you have just finished reading *"Wind in the Willows"*. Frogs and toads are quite different.

5.2.8 *Check your writing style*

To see how effective your beautiful prose actually is, read it aloud. This will tell you if the text "flows". If it doesn't, check that you have included a subject, a verb and an object in every sentence. Alternatively, see whether you are using punctuation appropriately. Reading aloud will enable you to see where punctuation marks *ought* to be placed.

Correct use of basic grammar includes the following:

- The appropriate use of commas, to indicate a short break in your reading. You should not use commas to join together ideas that are unrelated grammatically.
- The adoption of semi-colons to indicate a longer break; these are useful where related topics are contained within a sentence and their relationship could otherwise be joined by "but" or "and". Typically, the semi-colon is used to separate a main clause from one of lesser importance. A colon is used to separate a clause of text from a list.
- The use of full stops to indicate a significant break.

The following paragraph illustrates appropriate usage of commas, semi-colons, colons and full stops.

> The human body can gain or lose heat from its environment by four physical means: conduction, convection, radiation and evaporation. Normal cellular activity is possible only within a relatively constant temperature range; above this, proteins become irreversibly denatured, resulting in the deterioration in nerve and metabolic function. Temperatures much below the range result in a fatal slowing of metabolism.

5.2.9 *Some common errors of English usage*

Take care with "it's" and "its". "It's" is a contraction of "it is", as in "It's incorrect to use contractions in formal text". Do not use an apostrophe when the meaning is "of it" (as in "belonging to"), for example "…botulinum toxin is a neurotoxin: its main effects are….".

> …*it is* possible that a mammal may survive…
> may become:
> …*It's* possible that a mammal may survive…

Note that this form of contraction is usually inappropriate in scientific material.

Other than in the contractions mentioned above, the apostrophe is solely used to form possessives. For example, "the density of cells" may also be written "the cells' density, but not "the cell's density" or "the cells's density".

Apostrophes are never used in plurals, or "plural's" as some would incorrectly say.

While on the subject of apostrophes, in British usage dates do not have apostrophes, as in "*Roger Bannister broke the four-minute mile record in the 1950s*". Be aware, however, that Americans do like the apostrophe "*Neil Armstrong walked on the Moon in the 1960's*" but you should avoid the temptation to follow this example unless you are writing specifically for an American audience.

- The sharp-eyed among you may have noticed the **use of the hyphen** in the paragraph above and near the start of this sentence. A hyphen is used to form a "compound adjective", where two words are used to describe something. No hyphen is needed when words are used together as a noun phrase. An example is *"Charles Darwin was a nineteenth-century naturalist"* has a hyphen, but *"Charles Darwin was a naturalist working in the nineteenth century"* does not.

- Be aware of the difference between **abbreviations and contractions**. *"Prof."* is an abbreviation of *"Prof*essor" and to indicate this, it is followed by a full stop. In contrast, *"Dr* is a contraction of *"Doctor"* (the first and last letters are the same as the full word) and as such does not require a full stop.

- **Drug names do not begin with a capital** letter unless they are trade names, for example – "androstenediol" is a drug, "Bolandiol" is a trade name.

- The word **"data"** is the plural of **"datum"**; hence, you should use "data are...." rather than "data is....". Other plural words are: bacteria (singular: bacterium), media (singular: medium), criteria (singular: criterion), phenomena (singular: phenomenon).

- **Do not confuse "dependent" with "dependant"**: the former is an adjective, the latter a noun. So, responses are "concentration-dependent", not "dependant". A dependant is someone who depends on someone else; young children are their parents' dependants.

- **Do not confuse "affect" with "effect"**: either can be used as a noun; either can be used as a verb. Each meaning is distinct. "To affect" as a verb is to aspire to or to cause something to happen, to pretend to. The noun "affect" is a term psychologists use to describe a feeling. In contrast, "to effect" is to bring about or to accomplish and an "effect" is the result of an action.

- **Do not confuse "uninterested" with "disinterested"**: increasingly, people are using "disinterested" to imply lack of concern or indifference. The correct word here is "uninterested"; "disinterested" means not influenced by interests, as in a disinterested third party.

- **Do not confuse "imply" and "infer"**: the former means to suggest something whereas the latter means to conclude or make a deduction from something. So, a set of results may imply that a process is occurring and you can infer from these data that the process is occurring.

- **Be aware of homophones** – words that sound the same but mean different things, like "there" "their", and "they're" (they are), "two", "to" and "too", and "great" and "grate". A biological example is "weather", "whether" and "wether" (a castrated ram).

- Errors are often seen in the **use of "lead"**: "lead" is the present tense of the verb, whereas the past tense is "led". "Lead" as a noun, however, is the chemical element (chemical symbol, Pb) or, possibly, the strip of leather used to stop your dog from escaping.

- **Latin words or phrases should be in italics**, as in "Experiments performed *in vivo* led to different conclusions from those indicated by *in vitro* experiments". (Note the use of "led".) This rule does not apply to abbreviations such as e.g. (*exampli gratia* – for example) and i.e. (*id est* – that is), etc.

- **Names of species.** These should be in italics. The genus name (first word) should have a capital letter but the species name (second word) should not, and you may abbreviate the first letter after the first inclusion. For example, *Homo sapiens*: this can thereafter be written as *H. sapiens*. Care must be taken to avoid confusion when more than one genus name may share a single letter abbreviation.

- **Some words are frequently misspelled**: occurred, protein, receive, noradrenaline, albumin, penicillin, parallel, pipette are common examples.

- Remember that **even spellcheckers can be fooled.** The one that comes with *Microsoft Word* does not like "*noradrenaline*", which, because it is an American program, it would suggest is written as "*noradrenalin*". From the section above, *Word* doesn't like "botulinum", "androstenediol", "Bolandiol", "Pb" and "est", either. These would appear as "outline", "no suggestion", "Boabdil", "Pub" and "set", if you were to accept some of the suggestions *Word* makes!

5.2.10 *Some general points of style and punctuation in electronic documents*

- Punctuation marks such as commas, full stops, semi-colons, etc. should follow the preceding word without a space.

- There should be no spaces between brackets and the words that they contain.

- Paragraphs should typically follow the modern office style, as in this text. Start a new paragraph at the beginning of the line (do not tab or indent). Leave a blank line between paragraphs. Do not start a new line for a new sentence unless the break is natural; that is, the previous sentence finishes at the end of the previous line. Only press the return key when you want to start a new paragraph.

- Be aware of 'non-breaking' spaces. They should be used when you want to prevent breaks round a line, for example, in references or for initials and names. It is especially useful when using Latin binomials such as *E. coli* where the genus name is nearly always assumed. Press Ctrl-Shift and the spacebar to achieve this in *Microsoft Word*. Pressing CTRL-SHIFT and the minus key will produce a non-breaking hyphen.

- Electronic documents can easily be edited so there is no excuse for poor presentation. If you are unsure how to produce specific types of formatting, try the "Help" facility in the software package that you are using. Alternatively, the internet is an excellent source of information and "How to..." sites and discussion rooms are common.

5.3 Illustrations and tables of data

Illustrative material is very helpful in conveying messages succinctly and it has been said that a picture is worth a thousand words. It would not, however, be sensible to try submitting one picture in place of the one thousand words that you may have been asked to write as part of your in-course assessment!

Contrary to what you may have been told previously, figures and tables act as very useful adjuncts to scientific essays. Figures and tables can make difficult concepts seem simple or can summarize material very economically and so a few well-chosen figures or tables will add considerable value to your essays. Some points to remember about illustrations are:

- do not duplicate the same information by including it in a table and a graph
- number each figure or table in your essay so that you can refer to it at the appropriate point in the text
- give your figures and/or tables a title which is meaningful and indicates why you have included the information
- include a key to define any symbols or abbreviations, or a further short section of description unless the figure is very simple
- provide a reference to the source of the material unless you generated the data for a table or constructed the figure yourself

Remember that including illustrations or photographs in a document will often increase the file size significantly. Sometimes, it may make the document so large that it cannot be sent as an email attachment or, in some cases, stored on portable storage devices.

5.4 Referencing your work

At Level 1, a reference list may comprise a simple list of source material that you have consulted when preparing your essay. At higher levels, you will need to be more extensive in your use of references so that by Level 3 your essays and reports should resemble material in the peer-reviewed scientific literature. You should take care with your reference section, as incorrect or inaccurate referencing in essays is penalized. This includes omission of references either from the body of the text or from the reference list, inaccurate citation of papers, and also the inclusion of references that have not been cited in the text.

A good reference list starts when you begin searching the literature relating to your research topic. Much of your literature searching will involve the use of one of a number of bibliographic databases, such as "Web of Science" or "Medline". Each has advantages and disadvantages but whichever tool you choose to find your references, it is important that you keep accurate records of the references that you find *at the time that you use them*, so that you and others may refer to them later. It can be very difficult, and frustrating, trying to track down papers later on. Remember, also, that the web is an ephemeral source of information. This can be a particular problem with pages from Wikipedia, where anyone can edit a page at any time. Keeping track of changes can be difficult. It is sensible to keep a copy of the information that you access on the web. This may be either electronically or as a print copy.

When compiling your own reference lists for major documents, it may be useful to consider using specialized reference management software such as *EndNote*.

5.4.1 *What information is included in a reference?*

You may find lists of "References" at the end of book chapters that contain references to all the papers referred to in the text. Alternatively, lists of "Additional reading" may be given; these lists are usually not papers actually cited in the text but are more general review-type articles that the authors consider useful further reading. In either case, the individual reference may look something like:

> Watson, J.D. & Crick, F.H.C. (1953) MOLECULAR STRUCTURE OF NUCLEIC ACIDS: A Structure for Deoxyribonucleic Acid. *Nature,* *171* (*4356*): *737–738.*

The precise way in which a reference is cited can vary considerably, but they all tend to conform to a general pattern. In the example above, the first item is the authors' names. In this case, the paper was written by two authors (Watson, J.D. & Crick, F.H.C) – their names may be familiar. In papers with several authors, sometimes only the first author's name is given followed by "*et al.*" short for *et alii*, meaning "and others".

Next is written the year of publication (1953), in this case given in full and in brackets, and this is followed by the title of the paper (MOLECULAR STRUCTURE OF NUCLEIC ACIDS: A Structure for Deoxyribonucleic Acid). After this comes the name of the journal in which the paper appears (in this case, *Nature*). This is sometimes given in full but is usually abbreviated. Some abbreviations can be fairly obscure, although a knowledge of the subject matter may provide clues to the full title of the journal. There is no single standard for the abbreviation of journal names but if you are in any doubt, you would be wise to conform to abbreviations used by the Institute for Scientific Information, and referred to by the Web of Science. Journal and book titles conventionally appear in italics.

The series of numbers (*171* (*4356*): *737–738*) following the journal name indicate, in order; the volume in which the paper appears, in this case volume 171; the issue number in the volume that carries the paper "(4356)" – this is frequently omitted from reference lists; and finally, the first and last pages of the paper (737–738). In this case, the paper is very short, but nonetheless significant. This just proves that it is true that size doesn't matter.

The example above shows a fairly full reference. This is a most useful form of reference presentation because the more information you have about a reference, the easier it is to find, even if a mistake, such as the transposition of numbers, is introduced. For some purposes, a reference may only comprise a journal title, volume number, and a reference to the first page of the paper. Such references can prove impossible to trace if a mistake creeps into the citation.

5.4.2 *Styles of referencing*

Many universities recommend the use of the "Harvard" style for referencing material and this is the one that is used most commonly to cite references in academic works. The Harvard style uses author and year of publication in brackets in the text, and then the full reference details in the reference list.

Despite its name, the Harvard style does not fall under the jurisdiction of any single institution. Consequently, there is no authority that maintains the rules of this style and variations have evolved over time with different people having different ideas of what constitutes **the** Harvard style. So...

> *...it is essential that you follow precisely the instructions given on referencing for EVERY piece of work that you submit.*

EndNote, a commercially available bibliographic software package, has over one hundred reference styles for Biosciences journals, a similar number for those journals devoted to Medicine and a further twenty-five styles for Immunology journals. This compares with just one style for Mathematics journals! So, when presenting academic work, it is ***essential*** that the appropriate instructions be consulted before starting to write a manuscript. Expect problems from assessors, reviewers or editors if you do not conform to the house style for references.

5.4.3 A (fictitious) reference list

To illustrate the use of references in text and a reference list, you will find below a fictitious reference list including examples, some real (Sambrook *et al.*, 2001), others not (Brewer *et al.*, 1927, Dracula, 1897a), of different information sources, based on the Instructions for Authors for "*Microbiology*". This is an arbitrary choice used simply for illustrative purposes, and, of course, you would NEVER use fictitious references to support your coursework.

Brewer, B., Stewer, J., Gurney, P., Davey, P., Whiddon, D., Hawke, H. *et al.* **(1927).** On the road from Widdecombe Fair. *J Geogr* **12,** 1–10.

Dracula, C. (1897a). Blood disorders; a source for concern? *Vox Sang* **34,** 4–8.

Dracula, C. (1897b). Blood-borne infection; a source for concern? *Haematol Bluttrans* **56,** 123–140.

Dracula, C. (1898). A review of blood disorders in the Transylvanian populace. *Haemophilia* **78,** 5–14.

Frankenstein D. R., Igor, N. M. & Other, A. N. (1818). On the regeneration of human tissues. *Nature Biotechnol* 9, 333–345.

Jekyll, D. R. (1888). Lycanthropes wolf their food: a prospective study. *Vet Surgery* **10,** 666-668.

Jekyll, D. R. & Hyde, M. R. (1886). The unfortunate case of Mr H. In *Ghoulies and Ghosties and Long-leggety Beasties and Things that Go Bump in the Night*, pp 1–25. Edited by A. Spectre. Cluj-Napoca: Nosferatu Press.

Leeds University Library (2014) Referencing. [online]. University of Leeds. Available from: http://library.leeds.ac.uk/skills-referencing [Last accessed 12th May 2014].

Sambrook, J., Fritsch, E. & Maniatis, T. (2001). *Molecular Cloning: a Laboratory Manual*, 3rd edn. Cold Spring Harbor, NY: Cold Spring Harbor Laboratory Press.

van Helsing, A. (1893). *A study of the blood disorders of the population of Transylvania*. PhD, University of Guttenberg.

The first reference is to a multi-author paper that could have appeared in *Journal of Geography*. Old Uncle Tom Cobley was the team leader but he was the seventh author on the paper and does not appear in the citation, other than as "*et al.*". Next, there is a reference to one of two papers published by Dr Dracula as sole author in 1897. This may have appeared in the journal "*Vox Sanguinis*". It is followed by a reference to Dracula's second paper of that year, which may have been found in "*Haematologie und Bluttransfusion*". Following this reference is a single paper by Dracula again, appearing in 1898, possibly in the journal "*Haemophilia*". The paper by the Frankenstein team that could have appeared in 1818 may be found in *Nature Biotechnology*. Dr Jekyll is one of two authors of a paper that may be seen in *Veterinary Surgery*. He also wrote a chapter in a book edited by Prof. Spectre and published by the Nosferatu Press, which is located in Cluj-Napoca, the capital of Transylvania. Sambrook *et al.* wrote the *Molecular Cloning* book, published by Cold Spring Harbor Laboratory Press, located in New York. This is a real reference, as is the online citation to the University of Leeds web page on citing references. Lastly, you are shown a fictitious PhD thesis from a university that does not exist and written by one Abraham van Helsing.

5.4.4 *Compiling a reference list*

References in your reference list should refer to the actual source material that you consulted, **not** to the URL of the database in which you found your reference.

References in the list should be ordered as follows:

- alphabetically by surname of the first authors
- then, for any given first author, single-authored papers come before jointly authored papers
- within these categories (single-authored by X, then jointly authored with X as first author), the order is chronological

For the journal *Microbiology*, references should be formatted as follows:

1. surnames (in bold font) come first; include a comma after the surname followed by the initials (in bold) of that author. Each initial should be followed by a full stop and a space, and
   i. with up to six authors in a reference, separate author names by a comma immediately after the full stop of the last initial and separate the last two author names by an ampersand, "&"
   ii. generally, if there are more than six authors contributing to a paper, all those appearing after the sixth author are listed as "*et al.*", although sometimes you may be required to cite every author

2. year of publication (bracketed and in bold) followed by a full stop
3. title of the article in regular font, but using italics where appropriate for Latin names, etc.
4. title of the source of the citation, with titles of journals abbreviated according to the Institute for Scientific Information

Additional information and different formats may be required when referencing specific types of source (journal, book, thesis or internet) as shown in the following sections.

Difficulties with listing all the authors arise for papers to which a very large number of authors have contributed, such as the following, with 976 authors (imagine entering all those into a reference list!):

Van de Werf, F., Adgey, J., Ardissino, D., Armstrong, P. W., Aylward, P., Barbash, G., *et al.* **(1999).** *Single-bolus tenecteplase compared with front-loaded alteplase in acute myocardial infarction: the ASSENT-2 double-blind randomised trial. Lancet,* **354,** *716–722.*

References to journal articles

Abbreviated journal titles are given in italics **without full stops**; one-word titles are given in full. This is followed by the volume number of the journal (in bold), and then immediately by a comma. The citation is completed by citing the first and last pages of the paper, separated by a hyphen.

References to books

Following on from the title, these should also include the edition, editor(s) (if any), town of publication and publisher, in that order. When the reference is to a particular part of a book, the inclusive page numbers and the chapter title must be given.

References to a thesis

This is sometimes necessary. The citation should appear as follows: family name of the author, initial(s), (year), title, type of qualification, academic institution granting the degree.

References to electronic resources

As yet, there is no standard method, so the recommendations below are intended as guidance for those needing to cite such sources of information now. Documents on internet sources can be identified by the URL (uniform resource locator) or internet address. **Note:** with long URLs be careful to split only after the forward slashes in the address, and do not add further punctuation, such as hyphens, or alter the case of any characters in the address.

Give as much of the following information as is available: author/editor, year, title [online], (edition), place of publication, publisher (if ascertainable), available from: URL [date of last access].

If no author is specified, you should ascribe authorship to the smallest identifiable organizational unit (the standard method for citing works produced by a corporate body).

If the date of publication is not available, write "No date". It may be possible, in some cases, to find the date on which the page was created when using many web browsers, by right-clicking on a page and choosing "Properties" from the menu.

The term "online" in square brackets indicates the "type of medium" and is used for all internet sources. The term publisher is used here to cover both the traditional idea of a publisher of printed sources, as well as organizations responsible for maintaining sites on the internet. The "date of last access" is the date on which you viewed or downloaded the document. This allows for any subsequent modifications to the document common with this medium of communication.

Occasionally you may need to cite other sources of information. You should seek advice from your tutors or library team on such citations.

5.4.5 Citing references in your text

When citing a paper by **a single author**, your reference should consist of the author's surname, followed by a comma and then the date that the paper was published: for example (Dracula, 1897). If a paper has **two authors,** use "&" to join the authors rather than "and". Thus, you should write (Jekyll & Hyde, 1886) **not** (Jekyll and Hyde, 1886). For papers with **more than two authors**, use "*et al.*", as in (Frankenstein *et al.*, 1818). Note that "*et al.*" ends with a full stop, to denote that it is an abbreviation.

References to **different papers by the same author(s) in the same year** should be distinguished in the text and the reference list by the letters a, b, etc. Suppose Dracula was prolific in 1897. His works may be cited in the text of a report as (Dracula, 1897a) and (Dracula, 1897b) or, if citing two references together (Dracula, 1897a, b). If you need to cite a **group of references together**, it is recommended that you cite the earliest reference first. If you have **more than one citation for a given year**, then references should be cited in alphabetical order for that year. Each reference in a group should be separated with a semi-colon. If all the references used in the examples above were to be cited together, the result would look like this: (Frankenstein *et al.*, 1818; Jekyll & Hyde, 1886; Dracula, 1897a, b).

5.5 How to write a good essay

5.5.1 General considerations

Imagine that you are marking essays. If you were confronted with a large pile of scripts to mark what would **you** welcome most, and what would you hate?

A well-structured essay that shows clarity of thought and is legible is a joy to mark. Essays that are full of waffle, and that do not address their subjects are a real turn-off. How can you make sure your work is in the welcome category rather than being the most hated?

Any well-written essay or report requires information to be delivered in a well-defined structure with a beginning, a middle and an end (see Table 5.1): poor essays from inexperienced students often lack the beginning and end sections.

Table 5.1 Anatomy of an essay

The beginning or "the introduction"; – this places the subject of the essay in its context	Here, you define the subject of the essay. You do this so your reader knows whether to read on or not. Try to engage your reader with your topic to "persuade" him or her to continue, particularly if you are hoping to have your essay marked. Consider for whom the essay is being written. Do they have sufficient specialist knowledge to allow you to assume that they already have considerable insights into the topic? Or is your readership drawn from a broad background where you will need to include sufficient background material to allow them to follow your arguments? This is where you need to define the terms of your essay and explain your strategy. For example, you may give a brief overview of a broad topic in the introduction, reserving the in-depth exploration of selected examples to the body of your essay.

The middle or "the body"; – this is where you address the subject of your essay and develop ideas	This contains a series of linked arguments, each of which may have an introduction, a body and a summary. Depending on the length of the essay, you may consider splitting the body into different sections:
	
	    Point 1
	        Theme 1.1
	        Theme 1.2
	    Point 2
	        Theme 2.1
	    etc…
	
	Even if you don't make formal subdivisions in your essay, it is likely that you will break the body of your work naturally. At the simplest level, you will divide your text into paragraphs, each of which should be a self-contained unit expressing one idea.
	
	If the essay title was in the form of a question, **make sure that you answer it in this section of the essay.**

The end or "conclusion"; – this is where you should summarize the topic	This is where you make explicit the points you consider to be of most importance. Never introduce into your conclusions new material or ideas that are not covered previously in the essay. This is also a place where you should highlight any areas of uncertainty regarding the topic. You may wish to make suggestions as to how, in future, these uncertainties may be resolved.

5.5.2 *What is found in a good essay*

In broad terms, this classification can be related to Honours Degrees. Below, you will find a rough guide to what university tutors look for in answers of different classes.

First class answers

Students who gain first class marks take an "extended abstract" approach to writing essays. Here, topics in the essay are not only related to one another in a cohesive structure but, when relevant, they are related to wider body of knowledge and show an understanding of "the question behind the question". Students who attain first class marks link related topics and analyse information in depth.

Upper second class answers

Upper second class marks are the reward of those who apply a "relational" method to their writing. In this case, relevant topics are related to one another in an overall structure and are linked to form a cohesive argument, rather than being simply listed in text.

Lower second class answers

People who do not often progress beyond a basic approach to the set topic generally obtain lower second class marks. Here, ideas are considered independently in the essay, rather than being linked with one another.

Third class answers

Third class essays are written by students who just do not engage properly with the topic. Their work tends to be superficial, poorly organized, inaccurate and missing essential details, in addition to showing an absence of appropriate links between topics. There may be major errors and/or omissions from the text and irrelevant material is often present. Poorly worded and ungrammatical essays will also attain lower class marks.

Poor answers

These arise for two basic reasons.

1. Sometimes the writer does not understand the topic, or what the title is asking for, and so finds it impossible to put the right information together in a logical order.
2. Other times, the student simply hasn't spent enough time researching the topic and thinking about what information is required to address the title appropriately, or how to put the information together in a logical way. Starting your essay the night before it is due to be handed in is often the cause of the problem!

5.5.3 *Assessing essays*

How would you assess these short essays? If you were asked to mark them, to which of the three versions of a short essay devoted to cholesterol would you give the highest marks?

Version 1

A major ampipathic polar lipid, cholesterol, possesses both hydrophobic and hydrophilic regions. Cholesterol is an essential steroid lipid substance present in varying degrees in virtually every animal membranes constituting approximately 25% of the membrane lipids in certain nerve cells of all animal membranes. The great importance of cholesterols presence is it's capability of transportation of fatty acids, the metabolism of fatty acids and the production of hormones such as oestrogen and progesterone formation of vitamin D for the skin and formation of bile acids for food digestion. Nevertheless, for the insulation of nerve fibre's from damage. Conversely, excess cholesterol can cause diseases and death by depositing in certain areas of the arteries causing blockage leading to health problems such as heart diseases.

For the above advantages and disadvantages of cholesterol, cholesterol exists in the blood as two types, high density lipoproteins and low density lipoprotein. Both these lipoproteins differ in their densities upon which transport is based. Their combinations of lipids and proteins present vary depending on the particular lipoprotein, though they do similarize in their structure of a hydrophobic shell of proteins and a hydrophobic lipid core and amphipathic lipids. It is according to their lipid content that their role is determined, for example, HDL consists merely of a greater proportion of protein than lipids, therefore rich in phospholipids and cholesterol.

Version 2

To the public, cholesterol is probably the most famous and well-known lipid. It had long been accused by health professionals as the prominent cause of human cardiovascular diseases. Non-health professionals all think of it as a bad thing. But is it? To many health conscious people who didn't bother to check their facts, it may come as a shock to know that cholesterol has a crucial role in the structures of cell membranes and is a precursor of steroid hormones, bile acids and even vitamin D. The "good side" of cholesterol has not been well advertised at all. Therefore we should take time to explore this important lipid – is it really the culprit guilty of causing the death of millions of people worldwide?

To produce a balanced discussion, we will look first at what the body does with its cholesterol to see where things might go wrong. Cholesterol is not soluble in water, so the body uses "packages" of special types of proteins and fats called lipoproteins, to carry it around the blood. There are many different types, such as low-density lipoproteins (LDL), high-density lipoproteins (HDL), very low-density lipoproteins (VLDL) etc. which carry different amounts of cholesterol. LDLs carry the highest proportion of cholesterol and are considered "bad" by doctors since they are associated with heart disease and stroke. On the other hand, HDLs have a low proportion of cholesterol and are the "good guys" of the lipoprotein world.

Version 3

Cholesterol is an essential component of virtually every animal membrane, and comprises up to 25% of lipid in some nerve membranes. It is also important in the transportation and metabolism of fatty acids, in the production of hormones such as oestrogen, progesterone, in the formation of Vitamin D in the skin and in the formation of bile acids, which aid the digestion of food. Although cholesterol is essential, excess quantities can lead to health problems such as heart disease and in severe cases, death. The reasons why cholesterol, which is such an essential component of eukaryotic cells, can become a 'killer' will be explored in the following essay.

Cholesterol is carried in the blood as a constituent of high- and low-density lipoproteins (HDLs and LDLs respectively). The lipoproteins differ in their composition (which determines their densities) and they have a common structure of a hydrophobic lipid core and a hydrophilic shell. The higher density of the HDLs compared to LDLs is due to a greater proportion of proteins, phospholipids and cholesterol.

Did you note that Version 1 is full of grammatical errors? The second sentence talks of "every animal membranes", where "every animal membrane" is what is required. The next sentence has "it's" where the meaning is the possessive "its". "Nevertheless, for the insulation of nerve fibre's from damage" doesn't contain a verb and so is not a sentence. And things go downhill from there.

You may have found that the tabloid newspaper style and repeated questions fired at the reader in Version 2 is better but they make this a challenging read. Version 3, which avoids flashy style tricks and, consequently, appears authoritative is more likely to score best.

Checklist for writing good essays

- A good essay has a beginning, a middle, and an end.
- Figures and/or tables are welcome in essays but they need to be numbered unambiguously in the body of the text.
- When figures and/or tables are included, they should have a title; ideally they should also have a short explanatory legend.
- Scientific writing should be in the "third person".
- Try to achieve clarity of style; do not use long words and complex sentences designed to impress as very often they have the opposite effect.
- Avoid waffle in your essay.
- Stick to the subject set, rather than the topic you want to write about that sounds like it may be close.
- Read your work out loud and re-write the text until it "flows".
- *AVOID PLAGIARISM* – passing off someone else's work as your own is dishonest and carries severe penalties, culminating in exclusion from university.

5.5.4 Planning your essay

This is a very personal activity, and you may well find the following suggestions are not appropriate for you. We do, however, offer the following steps as one model for writing successful essays.

Steps in planning your essay

Step 1

- Obtain the title.
- Gather your thoughts on the topic and make notes.
- Prepare a rough outline of your ideas.
- Perform your literature search.

Step 2

- Prepare the first draft of your essay.

Step 3

- PAUSE and re-read the first draft.

Step 4

- Now write your second draft.
- PAUSE AGAIN and consider asking a friend for constructive criticism.

Step 5

- You are now ready to write your essay.

This should show you why it always takes longer to write an essay than you think it will. Let's explore the steps in more detail.

Step 1

Consider the title carefully and then jot down a few notes on the areas that you want to cover.

> *"Discuss the statement that the three-hour written examination paper is something that most graduates will not face again and fails to test skills necessary for a professional scientist. How best should these skills be assessed to permit determination of the degree class of a science graduate if not through traditional examinations?"*

If you were faced with this question, you should jot down your initial ideas. The question assumes that all graduates will pursue a career in science – is this valid? Practical training is under threat, and it may be argued that students are already doing too little at the cutting edge to allow for a meaningful assessment of these skills that could contribute to degree classification. This guide has already identified the areas that are considered important for graduates: personal skills, problem-solving and critical analysis, retrieval and handling of information, communication skills, teamwork.

Your essay may address some or all of these topics. Before long, you will have a number of headings that will form the body of your essay.

You can generate your notes in different ways.

- You could jot down what you know about a topic and then think about what information you consider to be the most important. From this, you could draw up a list of ideas, ranked in order of importance. These rankings then form the subheadings of the essay.

- Alternatively, you could produce a "mind map", a "spider diagram" or "spray", which are excellent ways of brainstorming a topic. To produce your mind map, start with the BIG IDEA in the centre of a blank piece of paper. Think of the topics that relate to the BIG IDEA and arrange these around the central figure. Some of these may lead to subtopics and to even more ideas. If you link your ideas with lines, you quickly build up a picture that looks like a web – hence the use of the term "spider diagrams" (see Figure 5.1). You do not need to restrict yourself to words – pictures can be incorporated into your diagram.

Figure 5.1 How ideas link in an essay

There is software available to help you to generate your mind map which allows you to play around with the arrangement of ideas very easily. In Figure 5.2, the subtopics have become sections of the essay and the rearrangement of the different branches makes it look much more like an essay layout. It is important that *you have a good idea of the outline of the essay before you start an in-depth literature search*. If you do not do this, then you may waste considerable time reading irrelevant material.

The internet is a frequently used source of initial information but it **must be used with great care** as explained earlier in this chapter. Having read your initial reference material, you may discover that you have not yet covered everything necessary for your essay, or you may decide that you have given greater prominence to particular ideas than they really need. At this stage, it is a relatively simple task to adjust the balance of your essay plan.

Step 2

Draft the essay. Find a quiet hour or so to write out a draft of the essay, at one sitting if possible, without too much detail at this stage. This will illustrate what you will need in your final introduction, body and summary. When writing this draft, you should bear in mind the suggestions on style and the use of good English, as set out previously.

Step 3

Re-read the draft, perhaps after leaving it aside for a day or two. Is it logical and are the ideas inter-related? Have you kept to the topic? Have you said what you want to say? If the answers are yes, yes and yes then you are ready to go on to the next stage.

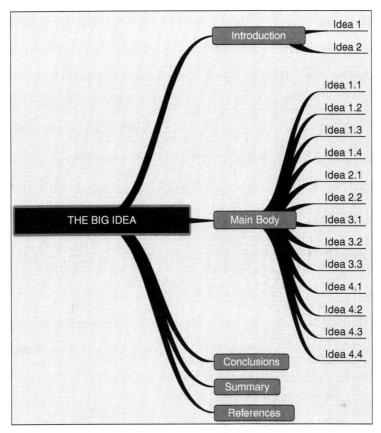

Figure 5.2 How linked ideas in an essay may become ordered

Step 4

Prepare a second, fuller draft – if there is time, put this second draft aside for a week or so, before the final stage.

Step 5

Prepare the final paper – if you are feeling brave, then you might like to ask a friend for constructive criticism. This is a useful exercise to ensure that you are really saying what you think you are saying. Check the spelling and grammar, and when you are finally satisfied with your effort, then you are ready to submit your essay.

Step 6

Submit the essay. You should only hand in an essay when you are convinced that it is the best that you can achieve in the time available to you. It is essential not to delete the working file of your essay, or any other work, until after the work has been marked and returned, and the module has concluded after the marks have been published. We are all human, and markers do lose work occasionally.

5.6 How to write a practical report

At some point, you will be required to write a report of your practical work. The style of report that is required is an amalgam of those used for the principal journals devoted to the publication of scientific research and it is essential that you follow this style. Your report is intended to communicate your results and conclusions, so it is very important to pay attention to the use of good English. It should also be presented clearly, neatly and with precision. You may need to present your reports using a word processor. Although this will give you a basic level of neatness, paying attention to the layout will bring dividends. Often you will be required to show the results of mathematical calculations. Except for the simplest of calculations, it is essential to show your workings. Unless you do, you will inevitably be penalized if you make a mistake. If you show your workings it is easier for the person who is marking your work to see where you went wrong and to provide you with appropriate feedback so that you will avoid the problem in future. Get into the habit of making "guesstimates" of the result that you may expect. Checking your graphs and results against your "guesstimate" will show if the results you wish to present are sensible.

Your report should be written in the third person, past tense. It should comprise a title page, introduction, methods, results, discussion, and references.

Common problems that you should avoid in practical reports

- A lack of organization such that information is not presented in a logical order.
- Disjointedness: each section should lead naturally into the next section.
- A lack of focus such that there is too much irrelevant information or not enough relevant information.
- A lack of critical rigour.
- Accepting all published papers at face value; just because something is published in peer-reviewed literature it need not necessarily be correct. There may be other papers presenting opposing views with equally convincing arguments, so you should always retain a critical view of what you read.
- Poor use of references; use of too many references that are too old to be of value; references that are not cited accurately; excessive reliance on reviews; references that are biased towards one research group; citing too few papers to support the arguments being expounded.
- Being too long!

5.6.1 The title page

This should begin with the title of your investigation, which should be appropriate to the work in the report and descriptive of the experiment that you have carried out. You must also include a unique identifier such as your name or student ID number on all paper copies of work for assessment, so that the mark may be attributed to you and the work returned after marking.

5.6.2 The introduction

This should include a description of the background information that relates to your experiment(s), written in your own words and including material beyond that provided in the practical schedule, when relevant. You should conclude the introduction by stating the aims of the experiment.

> **Checklist for a good introduction**
>
> • Give essential background information.
> • Set the scene.
> • Put forward the aims of the study.
> • Be succinct.

5.6.3 The methods section

When included, this should contain a concise description of the methods that you used, with sufficient detail to allow an experienced scientist to repeat your experiment without having to consult other sources. You will not always be required to include a methods section. For some exercises, this section is omitted entirely: on other occasions, you will only need to describe changes to the method provided for that particular exercise. Remember, however, that it is your responsibility to check what is required for each report. As with the introduction, this section should be written in your own words rather than being copied from the practical schedule that you were given to carry out the experiment. Try to avoid using too many linking phrases, such as *"and then we did this... and then we did that"* or *"...following this..."* or *"...following that..."*. Another common phrase in poor reports is *"...and then this was done..."*. A better way of phrasing material may be: *"Recordings were made after the light intensity in the bat cave had been adjusted"*. The methods section should be written as descriptive text rather than as a list or recipe.

5.6.4 The results section

This must be a text-based description of your observations that will include appropriate figures and/or tables. Your investigation may produce large quantities of "raw" data that will need to be refined before presentation in a practical report, and you will need to perform simple calculations to derive a useful set of data. For example, you may need to convert absorbance readings from an enzyme experiment into molar rate of change in concentration. Sometimes, when you first encounter a set of calculations it can appear that what is required of you is complicated and confusing. The more you practise, however, the easier calculations will become. It is a good idea to review all of the information that you have to carry out the required calculation.

Do not start your results section with a table or figure. In the early part of your undergraduate studies, you may be asked to provide data in the form of both figures and tables to check that you can draw accurate figures from your data; however, unless you are asked specifically for them, you should not present data twice. This convention is used for scientific papers and is good practice in scientific communication.

Unless you are describing complex pieces of work, where the results from one experiment lead on to the design of further investigations, you should not draw any conclusions in the results section of your report.

Presentation of results in a table

Raw data are often refined before they are presented in tables. When you use more than one table or figure, you should name them using sequential numbers (Table 1, Table 2, Table 3, etc.) and refer to them each in the text so that your readers can locate the correct figure or table with ease. Tables should include a short legend and should be presented neatly and, depending on the data that you are presenting, without gridlines. In Figure 5.3, which is easier to interpret: Table X or Table Y?

Table X. Presentation of results within a grid	
Classification	Result
1	X units
2	Y units
3	Z units

Table Y. Presentation of results without using a grid	
Classification	Result
1	X units
2	Y units
3	Z units

Figure 5.3 Presentation of data in a table with or without gridlines

With complex data and large tables, gridlines may guide the reader through the mass of information. With simple tables, however, reducing the lines in a grid makes the data more accessible. To see the effect of gridlines, look through papers published in the peer-reviewed scientific literature and see if you can find any presented in a full grid.

When appropriate, you should present the results of any statistical work that you have carried out to analyse your results. This should be descriptive and embedded in the body of the text. Do not be tempted just to cut and paste the results from your favourite statistics software package. If nothing else, by describing your results you will gain insight into what they mean.

Effective use of results

Having obtained your set of data, it is important that the best use should be made of the results. There are many types of figure from which you should choose the most appropriate format for the best presentation of your data: histograms, line graphs, scatter plots, pie charts, box plots, micrographs, diagrams, traces from instruments, etc. The most useful charts are likely to be line, bar or column graphs and scatter charts. However, pie charts can be effective, for example, in the presentation of the composition of fluids or populations of species in different habitats. **Be aware that for some reports, computer-generated plots are not always appropriate and hand-drawn plots are the preferred method of presentation of data.** *Catch Up Maths & Stats* gives more detailed advice on how to analyse and present data.

A note about units

Remember that all observations must be expressed together with their proper units. In a calculation this is a very useful way of ensuring that you have performed the calculation correctly because, if you are unsure which terms would be divided or multiplied by each other, the units can often be a guide. Units should combine or cancel out to give final terms that are appropriate to the required answer regardless of the numerical calculations involved. By forgetting to include appropriate units in a calculation, you may make errors and have no easy way of checking at which step you have gone wrong.

All figures should be numbered and include a legend in which the salient features are described succinctly. When appropriate, you should provide a key to the elements of the figure. The key may be included in the legend or may be part of the figure itself.

Independent and dependent variables

If a graph is the most appropriate form to represent your data, the "independent" variable should be plotted on the x-axis (horizontal or *abscissa*); with the "dependent" variable on the y-axis (vertical or *ordinate*). It is important to show all data points on graphs and it may be necessary to include error bars to indicate the spread of your results. The axes of your graph must be labelled to include the units in which your observations were measured.

Which is more useful, a curve or a straight line?

The two graphs below (Figure 5.4 and Figure 5.5) show the same data represented in different ways. Both show that the population grows with increasing rapidity as time progresses. Trying to read the population size accurately at the later times in the first graph becomes difficult due to the increasing slope. Therefore, the most useful graph is one in which the data points lie on a straight line, i.e. Figure 5.5. The data here have been "transformed" into a straight line by plotting the logarithm of the population size against time:

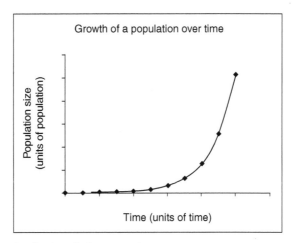

Figure 5.4 Growth of a population over time.
These data have been plotted without transformation.

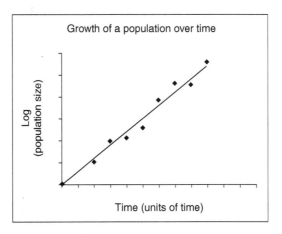

Figure 5.5 Growth of a population over time.
These data have been plotted after converting numbers of individuals in a
population to the logarithm of the number of individuals.

The best fit line

Any experimental observation is prone to error and so it is very unlikely that all of your
data points will lie on a straight line for any experiment that you ever perform. Your
points are much more likely to be scattered on either side of the "line of best fit".

The "best" straight line may not pass through any of the data points. Drawing the
"best" straight line on any graph takes practice and experience. As a rough guide,
however, the "best" line will have as many points above the line as fall below it
(assuming you have an even number of observations, of course). In Figure 5.4, the line
passes through the origin of the graph. but do not assume that all results will conform

to this. You should decide if it is appropriate for your line of best fit to pass through the origin or not. Also, remember that lines of best fit are not necessarily straight lines!

Handling data using Microsoft Excel

In some instances you may be able to use computer software to help with data handling. *Microsoft Excel* can be invaluable for handling repetitive calculations, for organizing data, and for presenting data as charts and graphs; it can also carry out some statistical analysis. Data are arranged in a spreadsheet with each cell (i.e. box in the grid) containing a single numerical value, or some text or a formula. Calculations can be performed on whole rows or columns of numbers simultaneously, saving many hours of laborious work. A big bonus is that if the value of any of the numbers in the spreadsheet changes, all the calculations involving that number will automatically be altered. By using *Excel* efficiently you can save a great deal of time and are likely to make fewer numerical errors when handling calculations involving large sets of data. Moreover, *Excel* allows you to express your data in a variety of chart formats.

If you produce and format tables in *Microsoft Excel* you can then transfer them straight into *Microsoft Word*. Additionally, it is possible to move data from a *Word* document into *Excel* for further processing.

Trendlines

The line of best fit can be calculated using "regression analysis", a very useful statistical tool which can also fit non-linear functions. To illustrate this, Figure 5.6 shows the line of best fit for a drug bioassay. In this experiment, samples containing antibiotic were placed in wells in an agar plate that was pre-seeded with a susceptible test organism. The more antibiotic present in a sample, the bigger was the zone of inhibition around the well containing the sample. In this graph, the

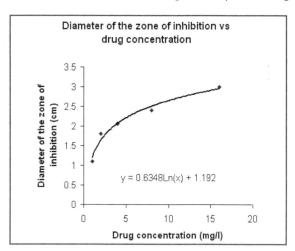

Figure 5.6 Line of best fit from a drug assay.
This figure is a screen shot from a *Microsoft Excel* spreadsheet used to find the trendline from this assay.

trendline does not pass through the origin but through the *y*-axis at the value of the diameter of the wells in which the samples were placed. The individual datum points were quite scattered and it can be seen that the trendline is curved. *Microsoft Excel* also calculated the equation of the line of best fit.

It is important that whenever you use standard curves to determine the value of an unknown sample, **any value for your "unknown" is interpolated from within the range of data points that you have obtained.** You should not be tempted to extrapolate your data beyond the lowest and highest values observed. There is no guarantee that the beautiful continuous curvilinear relationship seen in your data will, or should, extend beyond the limits of your observations.

Finally, you should always remember that there will be errors in your readings and you must take them into account when assessing any results you may obtain.

Checklist for a good results section

- Prepare figures and tables, if required, *before* writing the text.
- Plan the layout of the section.
- Present the results in a logical order, not necessarily in the order that the work was performed.
- Use section headings, if appropriate.
- Introduce each section, explaining why the work was considered important to the development of your ideas.
- Unless *specifically* told otherwise, use a figure *or* a table, but *not both*.
- Figures and tables must be self-contained and as such do not need to be confined to the results section – they may be used anywhere to enhance the clarity of the idea or to summarize complex information.
- When using figures you should:
  - ○ ask yourself what you really need to show
  - ○ show each individual data point on line graphs
  - ○ plot the "independent" variable – what you choose to observe – on the *x*-axis
  - ○ show error bars when appropriate, even on histograms
  - ○ avoid crowding too many separate graphs into one figure
  - ○ label axes and choose scales carefully
  - ○ remember to provide a key for each figure
- Only use illustrations if they are of good quality.
- Remember that illustrations may need special preparation.
- Be aware that there is no such thing as a "typical" gel or micrograph – these can take weeks of preparation!
- Remove surplus material from images but be careful to retain all useful information.
- Label illustrations carefully and clearly.

Here is another reminder: a numerical result requires the inclusion of appropriate units.

5.6.5 The discussion section

This is where you should discuss *all* of the results that you have presented in the previous section. Points that you may like to consider include:

- are the results accurate?
- are the results within the expected range of results for the experiment? If not, what might have gone wrong?
- do the results confirm or refute results from other experts? You should supply references to published information.

If the practical schedule from which you were working poses questions, the discussion section is the place where these should be addressed. You should highlight the significant findings of your experiment and discuss their importance. Your final paragraph should comprise the conclusions that you draw from your study.

You should finish your report with a reference list. References can be cited, both within the text and in the reference list, in a variety of ways and you should follow the guidelines given for your particular exercise carefully.

Checklist for a good discussion section

- Decide what constitutes your main findings.
- State clearly each of the main findings and discuss their relevance and implications.
- State how your findings compare with the work of others and cite references to relevant published papers.
- When necessary, refer back to previous sections of your report.
- Did you answer the question posed? If not then you will need to explain why not.
- What overall conclusions can you draw from your observations?
- What further work would you perform to follow up your observations or to support your argument?

What to avoid in your discussion

- Restating the results instead of interpreting them.
- Not putting your findings into a proper context.
- Repeating material presented in the introduction.
- Adopting a disorganized or illogical structure.
- Unclear and/or confused conclusions.

5.7 Posters

Posters are often used to communicate recent findings at conferences and other scientific meetings because they are one of the most efficient ways of presenting scientific research to a wider audience. Poster presentation sessions give authors the opportunity to discuss their work informally with other research workers and to pass on practical tips and techniques. Effective posters are those that explain the topic and the work undertaken clearly and are presented in a well-organized and logical manner. Guidelines for the style of presentation may differ according to the conference or course for which they are produced.

You should give careful thought to the presentation of your poster. A suggested format is shown in Figure 5.7. It will need a clear layout and should be readable at a distance. The basic structure requires an appropriate introduction, an outline of the work and a summary. Above all, it must make sense, giving reasons why you did the work and describing the conclusions that may be drawn from the results. The following points are worth bearing in mind.

- Remember that most people will probably spend only a few minutes reading your poster. You must get your message across in that time. Therefore, make sure your poster is succinct.
- Make use of short paragraphs or, preferably, bullet points rather than large blocks of text.
- Be ruthless in the material you discard. You do not have to report everything on your poster; you just need to tell a good story.
- Consider presenting your concepts and hypotheses as diagrams. Use figures or tables to summarize your results.

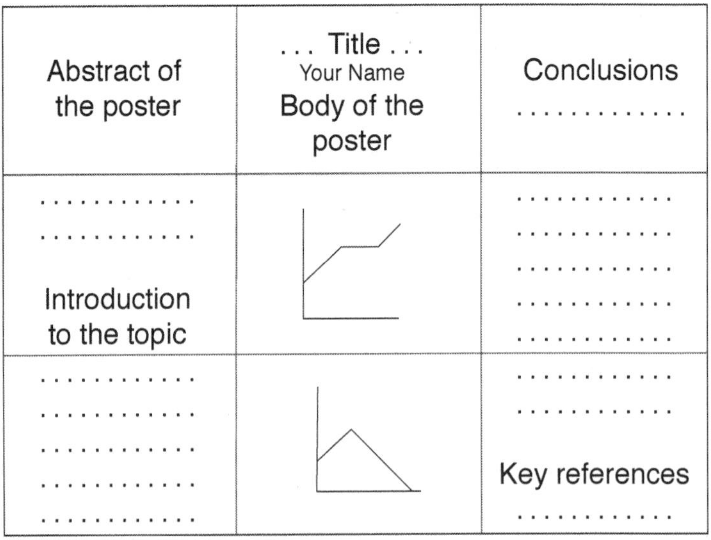

Figure 5.7 Suggested plan for a poster

To quote Jeremy Bentham:

> *"The more words there are, the more words there are about which doubts may be entertained".*

Checklist to help you make good poster presentations

- The poster should normally be produced in landscape format, unless you are told otherwise.
- Stick to the stated size limit.
- All text must be readable at a distance of 1 metre.
- Use at least 14 point text and preferably double space it.
- Photographs may be used and can be black and white or colour. If you do use photographs, make sure that you label them clearly.
- It is particularly important to remember to avoid large blocks of text on your poster.

5.8 Oral presentations

When first asked to make an oral presentation, many people feel apprehensive; this is perfectly normal. In this section, we give you advice that will help you to make effective oral presentations. Who knows, you may even end up enjoying them.

The key to a good oral presentation is ensuring that you are properly prepared.

5.8.1 Planning

When planning your presentation it is important to bear in mind the following points.

- What level of knowledge of your audience can you assume? At university, most of the talks you will give will be to your fellow students and you will have a pretty good idea of the level of understanding they will have. Later in your career, however, you may well be required to give talks to much broader audiences and you will need to decide how to pitch your talk. This can be quite an art.
- What is the purpose of the presentation?
- How long is the presentation meant to last?
- Ensure that you have a clear vision of the information that you want to communicate.

5.8.2 Resources for delivery

Make sure you know what facilities will be available to you when you make your oral presentation. You will look pretty silly if your presentation is created in a format incompatible with the equipment provided.

5.8.3 *Tips on using* **Microsoft PowerPoint**

These days, most presentations are made using *Microsoft PowerPoint*. While this is a very useful tool, it needs to be used with care. *PowerPoint* offers a bewildering array of backgrounds, most of which are horribly distracting; a minimalist approach often works best. Aim for a style and size of font that is simple and easy to read. Make sure that the colour of your text contrasts sufficiently with your background to enable the text to be read easily. Similarly, if you place images on a "busy" background, consider adding a border so that the image may be distinguished without difficulty. Cluttered diagrams and figures are a big turn-off for audiences.

PowerPoint has some wonderful gimmicks and animations. However, try not to use these over-enthusiastically. If every piece of text floats in and then pirouettes into its final resting place, your audience may well end up feeling mildly irritated or, worse, seasick. When it comes to animations, "*less is more*".

5.8.4 *Content and organization*

You are now ready to write your presentation. Like good essays, good oral presentations have three parts: a beginning, a middle, and an end (see Table 5.2).

Table 5.2 Structure of a good oral presentation

Beginning or "Introduction"	This is where you grab the attention of your audience and inform them of the subject of the presentation.
Middle or "Main body"	This is where you need to present your information in a logical order. Good visual aids are extremely important to help explain systems and concepts.
End or "Conclusion" or "Summing Up"	Here, you should summarize the main points of your presentation, referring back to your introduction if possible.

The advice on the structure of the presentation is the same as for a good lecture. To reiterate, you need to "...*tell them what you are going to tell them, you tell them, and then you tell them what you've just told them*"!

When designing your talk, it is worth remembering a few points.

- Assume that the information on an average *PowerPoint* slide will take about two minutes to deliver. So, assuming that you have an average amount of information, for a ten-minute talk, you should aim to have no more than six slides.
- Avoid information overload and so try not to have more than six lines of text per slide.
- Use short phrases rather than sentences so that the audience can focus on the main points.
- Avoid producing cluttered diagrams and figures.
- Check for spelling mistakes.

5.8.5 Practice

Having prepared your slides, your next task is to rehearse your talk. Try to get access to the room in which you will be making your presentation before the event so that you can get a feel for the ambience. Read your presentation aloud. This will highlight issues regarding timing and the sense of what you are trying to communicate. Although it may feel a little odd at first, rehearsing in conditions as close as possible to those for the real presentation will pay dividends.

> **Tip**
>
> Compile your "script" as a series of key points on index cards that act as aides-memoires. If you use more than one card for your script (one per key point?), it is a good idea to tie them together or number them in case you drop them during the talk itself.

What to avoid for a good presentation

Do not be tempted to try to learn your presentation by heart. If you do, it is likely that your delivery will be rather stilted or you will forget your lines. Also, don't use a full script and read everything from that. For one thing, if you are using a text-based presenting tool like *PowerPoint*, your audience can read the key words faster than you can deliver them and, if your head is buried in your notes, your voice will not project to the back of the room. Worst of all, if you lose your place in the script, there will follow an embarrassed silence whilst you try to find where you had got to on the sheet! Finally, you will annoy your audience considerably if you decide simply to read your slides.

5.8.6 Your final presentation

Try to remember the following suggestions when you deliver your oral presentation.

Face your audience and make eye contact as much as possible

But don't stare at any one individual. Do not be tempted to face the screen on which your talk is being projected since this alienates your listeners and your voice will not project out to the audience.

Use a pointer

This helps to draw attention to specific parts of a slide. Be careful, however; laser pointers should *never* be pointed at your audience. If your nerves get the better of you and you find the pointer wobbling about uncontrollably, why not use both hands to steady it?

Pace your delivery

A good talk is not delivered at the same pace as you talk in normal conversation – it is somewhat slower. Interestingly, when you are nervous, the temptation is to speed up your delivery rather than slow it down. It is a good idea to introduce deliberate pauses into your delivery. This may seem strange at first, but it is a technique that works. If you watch videos of people making speeches, you can often count three beats between sentences – longer if the audience responds to the last point being made. Also, remember to take regular, large breaths. It does wonders for voice projection but it is something that can be easily overlooked if you are feeling nervous when presenting to a large group, many of whom you do not know. Regular breathing also forces pauses in your delivery, which is a good thing.

Try to exude enthusiasm and confidence

This is commonplace amongst successful speakers. Don't panic if you find that you have accidentally skipped some information. If you are confident in your delivery, most of your audience won't notice, even if you have the relevant bullet point projected on screen. Confidence comes with knowing your topic. The better you know the subject about which you are talking, the more relaxed and comfortable you will be with your presentation. Remember, though, in almost any presentation that you will ever make, you will probably know more about the topic than most people in the audience.

5.8.7 Dealing with questions

After almost all oral presentations, your audience will have the opportunity to ask you questions. At first, this may seem daunting but you can prepare for this. While creating and constructing your presentation, think about the questions *you* would ask and make sure you have found the answers. If similar questions occur to your audience you are half way to answering them.

When you do get questions, answer the ones that have been asked, rather than trying to bend your answer to fit the question you would like to have been asked.

If you haven't heard the question properly, politely request the questioner to repeat it. If you don't know the answer, don't pretend that you do. You will only dig a hole for yourself and could end up looking seriously silly. If you do find that you are in a hole – stop digging!

Lastly, try to avoid foot-in-mouth disease. Think about what you are going to say before responding rather than spouting the first thing that comes into your head.

This was an experience of a postdoctoral student applying for a lecturing post, who avoided foot-in-mouth disease:

> "One of the interview panel was asking a long series of questions about the nature of mobile genetic elements when he suddenly asked me to name a third-generation cephalosporin. Given the series of questions that had gone before, I was somewhat taken aback. I paused rather than give a stupid answer. My interviewer barked back at me "Hurry up! They all begin with 'Cef-'". I continued my pause, and in my own time answered "ceftazidime" – the right answer; and I got the job."

5.8.8 Improving your presentation technique

Once you have finished your presentation you can heave a huge sigh of relief. If, however, you want to make the most of the session, ask a trusted friend for constructive feedback. There are aspects of every presentation that can be improved. If you are ever asked to give advice, make sure criticism is constructive. Think about which you would prefer to hear: "That talk was rubbish" or "That talk went right over our heads. Next time why don't you try doing…". Also, the more times you give oral presentations, the easier they become.

Checklist to help you make good oral presentations

- Prepare thoroughly.
- Know your topic. The better you understand the topic you are presenting, the easier it will be to make an effective talk.
- Make sure you know what equipment, if any, will be available in the venue where you will talk.
- Don't overcrowd information onto visual aids and don't use backgrounds that are distracting.
- Make sure any text on your visual aids is readable.
- Use pictures, they are invaluable in good presentations.
- Use index cards containing a few key words only as an aide-memoire.
- Remember that you can fit in far less information per unit time than you think.
- Be confident in your presentation; look at the audience and speak clearly and relatively slowly.
- As soon as possible after the presentation, get constructive feedback from a friend (and be prepared to do the same for them).

5.9 Summative and formative assessment

In educational jargon, "**summative assessments**" are those that contribute to your final module mark. Many courses also make use of "**formative assessments**" which are assessments that provide guidance for you. They help you to know how you are getting on in the course, but they do not count towards your final module mark.

Formative assessment may be given in different ways. For example, it may come from tutors and laboratory demonstrators as written feedback on an assignment; helpful feedback will praise the good aspects of your work while pointing out any shortcomings to help you to see how you may improve in future assignments. Alternatively, you may receive interactive feedback when using online resources. You will only understand what the feedback you receive means by looking carefully at the comments and at those aspects of your work to which they relate. You can then use this advice when producing future work. If you do not understand what the feedback means, it is important that you ask for clarification from the person who gave the feedback.

There is currently a move towards more self-directed learning. To ensure that your learning is appropriate, formative assessment will play an increasing role. Try to develop a systematic and productive approach to your feedback by using the checklist below; you might be surprised by how much your marks improve.

Checklist for making the most of feedback

- Keep a list of the most significant comments and what you understand these to mean.
- Recognize which aspects of your work are praised and make sure you keep doing these.
- Recognize which aspects of your work receive criticism and make sure you improve these in future work.
- Notice any recurring negative comments which might indicate that you do not understand how to correct problems that have been highlighted in previous work.
- Having done all this, you can then concentrate on the main aspects of your work that need improving while maintaining the things that you do well.

Chapter 6 Using computers and information technology

At university, you will have plenty of opportunities to practise your computing skills and will have access to a range of software to help you with your studies.

Microsoft Word and similar programs allow you to generate and organize documents. You can easily cut, paste and format your text, and mistakes are easy to rectify. Many have spell-checking facilities and some can even check your grammar. Remember, however, they are all fallible and they may try to suggest changes to your writing that are patently stupid!

Presentation – graphics software, such as *Microsoft PowerPoint*, is used to make dynamic presentations that are valuable support material for your oral presentations.

Spreadsheets, including *Microsoft Excel*, are used for storing and sorting numerical data. They also perform calculations automatically. Many spreadsheets such as *Excel* can convert numerical data into a graphical presentation.

Databases, such as those produced using *Microsoft Access*, are collections of records somewhat like an old-fashioned card index. Because the computer does most of the hard work for you, databases can be linked together to provide very powerful information storage and retrieval – far more than could be imagined using a card-based system. Whereas spreadsheets, such as those produced using *Excel*, are viewed as a 'flat' or two-dimensional form, databases allow you to relate information between different spreadsheets simultaneously in a multi-dimensional manner. This is useful when you want to pick out and compare specific items of information about all the subjects in a group for which you have recorded a range of details. When setting up a database it is important to have decided *at the start* every one of the characteristics you wish to survey, otherwise you will have to trawl through each of the individual spreadsheets later, in order to add the extra details required.

Statistical packages are designed to take the hard work out of statistical calculations. You will be taught which statistical tests to use when in lectures and practical classes.

Graphics packages are commonly used by scientists to create graphs, pie charts, etc. They take the pain out of data presentation. These days such programs are often embedded in other packages such as spreadsheets.

6.1 Interactive online learning

These days, traditional learning is augmented with a variety of interactive online materials. They may range from resources that deliver simple multiple-choice question and answer tests through to simulations and computer models used to illustrate complex concepts that are difficult to present in any other way. This computer-based learning (CBL) is sometimes known as computer-assisted learning (CAL) and is increasingly being delivered via the internet, meaning that you can access material from anywhere in the world. Another name for this is technology-enhanced learning, or TEL.

6.2 Being organized – file storage and management

It is important to get into the habit of being organized about storing your files with appropriate identifying names, and the files sorted into suitably ordered (and named) folders. Otherwise, you will very soon discover that keeping track of files becomes very hard, and that the document you really need to access, written just a couple of months previously, is lost in the morass of files that are in your folder. The Windows operating system allows you to do this, but beware the "default" save facility since this will keep adding files to the "My Documents" folder without placing the file into an appropriate folder. While you can choose any name for your folders, it is a good idea to give them logical names. Will you remember that you saved "mydoc1" or "coursework_4" in "the seventh circle of the seventh circle" weeks or months after you saved them? And will you remember what information each file contained? Figure 6.1 shows an example of a file structure that has a logical basis:

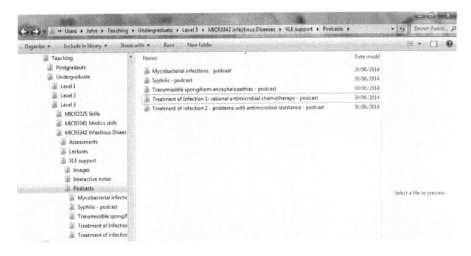

Figure 6.1 An example of the file structure on a computer

In this example you should note that the teaching folder contains separate folders for each level of study. In each will be sub-folders devoted to particular modules and these may contain sub-folders relating to teaching activities...

Jonathan Swift observed:

"So, nat'ralists observe, a flea
Hath smaller fleas that on him prey,
And these have smaller fleas that bite 'em,
And so proceed ad infinitum".

This has been expressed more poetically as:

"Great fleas have little fleas upon their backs to bite 'em,
And little fleas have lesser fleas, and so ad infinitum.
And the great fleas themselves, in turn, have greater fleas to go on,
While these again have greater still, and greater still, and so on"

So it is with file trees.

Organized file structures work best when you give your files logical names rather than relying on what your software assumes will be a sensible name – usually the first line of the text in the document.

6.3 Keep file sizes small

Although computer memory has probably never been so cheap and plentiful, you should occasionally pause to think about file sizes. Images cause a particular problem here. Using *Microsoft Paint*, the default file type is the "bitmap", where every dot in a picture has its position and colour recorded. This can generate enormous files, even for relatively small images. To overcome the problem of large file sizes for images, a number of other formats that compress your data have been developed. Among the most popular are "GIF" (short for *"Graphics Interchange Format"*) and the "JPG" or "JPEG" (derived from the name of the *"Joint Photographic Experts Group"*) file types. Both have advantages and disadvantages over each other (see Table 6.1). Another file format, the "PNG" or *"Portable Network Graphics"* file, overcomes many of the problems with GIF and JPG files, but it does not support animations.

6.4 Back-ups and passwords

Migration of files between computers has become commonplace. Portable memory devices to use for such transfers have become much more reliable in recent years; nevertheless, they, and hard drives, do occasionally fail. This can be an extremely traumatic experience. Cloud resources are becoming increasingly popular for the storage of electronic materials since they have the advantage that files can be accessed from anywhere with an internet connection. However, you should be aware of potential security issues. Regardless of how you choose to store files, it is probably a good idea to ensure that you keep backup copies in more than one location (such as an external hard drive, a CD-ROM or USB key) and that you coordinate copies of your archived material on the different storage devices that you choose to use.

Table 6.1 Comparison of image file types

	GIF	JPEG	PNG
Advantages	Good for saving line diagrams; can deliver animations	Good for complex images like photographs	Has file sizes intermediate between GIFs and JPEGs but with better colour depth
Disadvantages	Uses only 256 colours and so a high-resolution image may suffer from alterations in hue and some loss of definition	Suffers loss of definition around boundaries so text and clean lines may become blurred; the JPG format is not recommended for storing images that require subsequent editing since there is sequential loss of definition	Cannot display animations and not supported universally so some software packages cannot use this format

It is also not unknown for computers to hang (just as you have finished your finest piece of work, but before you have saved it) losing your work and hours of your time. To avoid this problem, make sure you save your document every 5–10 minutes. On your own computer you can set it to save documents automatically at defined intervals.

Many systems require a password before you can access the information you require. It is best not to use the same password for all these systems (access to an online bank account, academic resources and also your computer, for example), but if you have too many different ones you are liable to forget which ones apply to specific situations. To avoid this, you should keep a log of your passwords somewhere safe. Remember, you should use a mix of unusual characters including symbols and a mix of upper- and lower-case letters in passwords. If you can, devise your own logical system to create passwords that are unique to a particular facility or website but easily remembered by you. You could also consider using one of a number of online password management systems. If you store your password list in an electronic document, remember that this can be protected by using the security features of the program to impose a password that is required to open your document.

6.5 Email and etiquette

It is very likely that you already have multiple email addresses, used for different purposes. When you enrol at a university, however, you will be allocated a university email account. Always use this when communicating with your tutors or with anyone else when communicating about university issues. The one exception to this advice, however, is when you are applying for jobs late in your final year. Shortly following graduation, your university account will be shut down, cutting off an essential line of communication. One of the advantages of email is that you can respond instantaneously to someone. Indeed, you can hold a "real-time" conversation. The speed with which you can communicate, however, can also be a disadvantage. Email communication is never subtle and it is all too easy to send an email that you may later regret sending. This is a particular problem when sending messages about which you have strong feelings. Remember that you can always save a draft of an email and send it later, having read it carefully before committing yourself to pressing the "send" button. Think about what effect your message may have on the recipient and ask yourself if you would want to receive the rant that you are about to send. What does it say about you? Likewise, beware of "humorous" email addresses. You can change your email alias easily, but do you really want to advertise to the world that you are a "studmuffin" or a "sexy_babe"? These are both names that students have used. Remember, also, if you abuse university computing services, you will leave yourself open to disciplinary action being taken against you. In extreme cases, abuse has led to successful criminal prosecution.

6.5.1 *Sending emails*

Make sure you include a subject, so that the recipient knows what the mail is about when it arrives at the inbox. Some mailing programs will remind you to do this, but others don't.

You can send a message to more than one person by:

- including additional addresses in the 'To' box
- including addresses in the 'cc' box (carbon copy) – where addresses in the 'cc' box will appear on the recipients' email, or the 'bcc' box (blind carbon copy) which is useful if you don't want all the addresses to appear in the recipients' 'cc' box
- creating and using a distribution list containing email addresses of people you regularly want to contact (e.g. members of a committee) – this can save a lot of time

Creating a message

Emails may be word processed in much the same way as documents in *Word*, including cutting and pasting, dragging and dropping, making bold and italic, etc. Clicking the right-hand button on the mouse can be used to activate sub-menus of useful operations and you can insert hyperlinks to web pages, so recipients are taken straight to the page by clicking on the link.

Adding attachments

Use attachments to send large documents rather than trying to include them in the body of a standard mail message. Click on the attachments icon/box, and then browse for the file you require on the mini-file directory that appears. This facility is particularly useful for sending files such as coursework, CVs, minutes of meetings, pictures, and so on, to accompany a covering letter in the main email message. Sometimes it is important to share very large files, such as a report containing a large number of detailed images, and there is a limit to how much you can send via an email. Fortunately, there are very good alternatives these days, including WeTransfer, Dropbox and Skydrive.

> **Note**
>
> Try not to send large files as attachments. People are reluctant to clear out old emails and if too many emails with large attachments accumulate in your inbox, you will be in danger of exceeding your storage quota. When this happens, you will no longer be able to receive any emails. This can mean that you could miss vital messages. In summary, try to keep emails small.

6.5.2 Etiquette

Although you may be informal amongst friends who know you well, remember that when you address others who know you less well (tutors, lecturers, prospective employers) they will form an opinion of you from what you write. Good sense says that you should make the effort to write in a grammatical and (reasonably) formal way that shows some respect to the recipient, otherwise you will appear off-hand, sloppy and rude. The box below summarizes the points to remember.

What to do and not to do in emails	
Do: <ul><li>include a proper subject line to inform the recipient of the email contents</li><li>begin with: "*Dear Dr / Prof. / Mr,*" etc.</li><li>be polite and be grammatically correct, e.g. "*Thank you for contacting me about the meeting. I will be able to attend.*"</li><li>finish with your name (including surname). You might also want to include: "*Best wishes*" or "*Thank you*", or "*I look forward to your reply*"</li></ul>	Do not: <ul><li>forget to include a subject</li><li>begin with: "*Hi*", "*Hiya*", or even just launch into the main part of the message</li><li>use slang, text language or sloppy grammar, e.g. "*yeh, il b ther 4 the mting*".</li><li>forget to include your name at the end (or the recipient might not be able to identify who has sent the message), or include: "*Luv Cherie x*", or "*cya l8r*", or "*thanx*"</li></ul>

Email systems *usually* allow you to add a 'signature' to the end of your message – once set up this means that your title, position, address, phone number, etc. can be added automatically to each message you send.

6.5.3 *Replying to messages*

Be aware of the difference between the '*Reply to all*' and '*Reply*' buttons; the first will send your reply to all those in the 'cc' box AND the 'bcc' box, whereas the second will only reply to the sender. If your reply contains sensitive material, creating a new message to the sender alone might be the safer option. The message to which you reply will automatically appear in your reply but you can delete the non-essential parts of that message in order to save space.

6.5.4 *Organizing your emails*

Try to be tidy. Create folders within your mailer into which you can move related emails from your inbox (many systems allow you to do this simply by dragging and dropping). Delete unwanted emails so that your file space doesn't exceed any preset limit. If this happens, you may no longer be able to receive and send emails, which could have serious consequences if your department sends you information by email about important deadlines.

DO NOT OPEN ATTACHMENTS unless you are confident that they do not contain a virus. Always check:

- whether you know the sender – if you haven't any idea who they are, delete the file
- whether the subject looks sensible – ignore money-making offers, emails supposedly from banks or emails enticing you to augment parts of your anatomy! Even if these do not contain viruses, they are spam, they clog up the internet and should be deleted. Never reply to these because the organization will then 'retain' your email, send you more messages and sell your email address to other similar organizations.
- whether you were expecting an attachment – if you receive something odd, apparently from a friend, email them in a separate mail, to check that they really did send it
- the file format – those ending with .exe are executable files and may start a program running on your computer; they should be deleted immediately. Unfortunately file attachments with more common formats such as '.doc' may also contain viruses, so do check as per the first three points above before you open these files.

6.6 University printing facilities

An area where many students have traditionally experienced problems at some time during their degree is printing. While most people have their own printers for routine work these days, many still rely on university printers for high-quality productions, such as colour printing for final year dissertations. You should remember that specialist printing facilities on campus are finite and demand for them is not spread evenly. So, as the deadline for any assessment approaches, the length of printer queues grows with alarming rapidity and the time you have to wait to get your work printed gets longer and longer. It is also all too easy to send multiple copies of a file for printing, particularly when you are under pressure. This will cost you money and it will create problems for other printer users. The last week of any term is a particularly bad time for printer queues, but any deadline for a large module can cause problems. Don't assume that because you don't have a deadline that there won't be a wait to get your words of wisdom into hard copy. There is a simple solution to this problem: *plan to get the work completed before the deadline.*

6.7 "What to buy or not to buy; that is the question..."

When computers cost an arm and a leg and were as big as a walk-in fridge, a frequently asked question was "Will buying a computer help my studies?". There has been a dramatic drop in the cost of computers that has been matched with an enormous diversification of the number of ways in which to access online material. Smartphones now provide access to the internet and tablets are almost universal, so the question of *"Can I afford not to have a computer?"* is now redundant. A much more important question these days, however, and one to which there is no easy answer, is *"What type of device will I need?"*. Even within a single institution, different faculties may support different types of device, and resources that can be viewed on some devices will not be available to people who favour others. Before committing yourself to a particular device or operating system, try to find out whether it will be suitable to support your studies. Bear in mind, also, that there are an enormous number of games and other distractions available these days. While these have an important role in relaxation, it can be tempting to try to complete just one more level before revising and this can have serious consequences.

One final note of caution:

> *Computers can be used to achieve some wonderful work and they can be great fun. They can also become great time wasters. It is essential that you distinguish between computing for work and computing for pleasure!*

Chapter 7 Revision and examinations

To misquote the late Bill Shankly on football, *"Examinations are not a matter of life or death. They are much more important than that"*.

Many students doubtless recognize this statement. The run-up to the exam period and exams themselves may feel like a nightmare, filled with panic, stress, anxiety attacks, caffeine fixes and wandering around like a zombie. You may also wake up in the middle of the night in a cold sweat, convinced that you haven't done enough revision. There are ways in which you can avoid the worst excesses of exam stress. You just need to decide that you will work hard and that you can succeed. Everything else will then follow from this.

The keys to exam success are:

- good organization
- self-discipline
- hard work

7.1 Effective revision

7.1.1 Plan a realistic revision schedule

This should be done well in advance of the examination date and you should think carefully about what you are trying to achieve. Only you know how much you have to cover and you will quickly discover how much you can achieve in a day so it is sensible to start in plenty of time. This means preparing a timetable so that you anticipate having completed your revision a few days before the date of your first examination. You should leave yourself time to spare so that you can cope with the unexpected. Furthermore, when you do end up with this unstructured time at the end of your revision, you can use it to go over those topics with which you are least happy. You should:

- remember that examinations test your knowledge and understanding over the **whole** subject area so you need to cover **all** of the topics covered in a module
- ask yourself if you want to cover something once, or several times; planning your time also means building in flexibility and breaks
- write out a realistic timetable covering all the topics that you need to revise

Detailed knowledge of a small part of the course will not compensate for large areas of ignorance. Students who do poorly in module exams based on essay-style questions, often do so because, although they can tackle two questions with ease, they have not revised enough of the module to permit them to tackle a third topic adequately. This is painful and is best avoided.

7.1.2 Revise little and often

It is more effective and less stressful than cramming. You will also remember and understand more of a topic if you cover it three times in three separate one-hour slots than you will by spending three hours at one sitting on the same material. Remember: **spaced learning promotes retention and recall.**

7.1.3 Plan your method of revision

Active revision is much more effective than passively reading through your notes and will help you to avoid the temptation simply to stare at your notes and textbooks, fooling yourself into thinking you are doing something useful. If you have taken advice given earlier in this guide then, overall, your work will be reasonably organized and summarized. If this is not the case, then your first task is to organize and summarize the material. Then you can start your active revision by constructing your own set of revision notes. Read a section of your notes/handouts/textbook(s). Ask yourself which are the key facts and concepts. Now put the book away and see if you can jot the key points down. Look for connections to other topics covered in both lectures and practicals.

Revision cards

Many people find it useful to summarize and condense topics progressively until the key information can be fitted onto a postcard. One side can list the main headings and sub-headings, and the other, simple diagrams and brief notes. Don't forget to include links to other topics. A collection of these cards provides you with a portable revision pack that can be used in bus queues, on the train, etc., as well as during more formal revision times.

Spider diagrams or mind maps for revision

An alternative to the use of cards is to generate a spider diagram or mind map to help you to revise. You need not limit your map to words – pictures can be added and these provide a powerful tool when trying to remember key elements of the topic while you are under examination conditions. Figure 7.1 shows a picture of a revision mind map with pictures. The map has been shrunk deliberately so that you cannot see the details, which don't matter. The important thing is to see what a difference a few key images and the imaginative use of colour can make. OK, for the very curious, the image in this case was revision notes based on the life and works of the poet Byron, who was *"mad, bad and dangerous to know"*. Several computer programs can assist with this.

Test yourself with practice questions once you have revised a topic

For many examinations you can obtain copies of previous examination papers. These will give you a flavour of the type of questions that have been asked in the past. Tackling past examination papers will give you practice at writing examination answers. One of the best ways of doing this is to revise a topic, then attempt a past question under examination conditions. Allow yourself only 30 or 40 minutes to write down your answer, and do not refer to notes. When you have done this, look

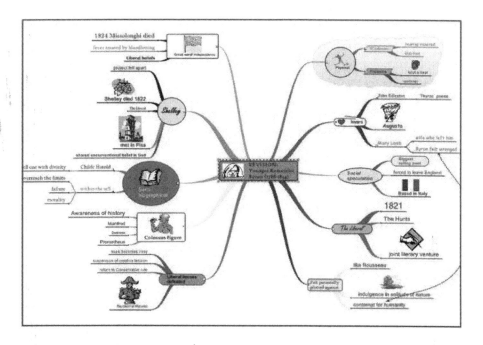

Figure 7.1 A mind map with pictures to act as a revision aid

at your answer, and see if it is structured well. You may like to refer back to "How to write a good essay" (*Section 5.5*) to judge this.

You should, however, treat past papers with a degree of caution. There may be questions that you could not possibly answer because courses evolve over time, and what was important in previous years may have been discarded to make room for new material. Furthermore, it may be that a member of staff who habitually used to ask questions on a particular topic is no longer working in the institution, and so this topic may no longer be taught in as much depth. Remember, however, great care is taken when setting examination papers to ensure that they are fair and that they cover the course at the time the examination is set. Your lecturers set the examinations, and have a detailed understanding of your curriculum so will set questions that should enable you to demonstrate your knowledge.

An exercise to try

Pick two questions that you consider you could answer equally well. Under time constraints typical for examinations (say 30 to 40 minutes) write out your answers to each. There is an additional rule. For one question you must start writing immediately. For the other you must NOT begin to write your essay for at least 5 minutes. During that time, however, you can formulate a plan of the outline of your essay. You may even like to make a spider diagram or mind map of things to include in your essay. Imagine that you had to mark your essays. Which approach produced the better answer? Which approach will you adopt in future?

Group revision

Some people find that working in pairs or in small groups saves a lot of hard slog. The work can be shared out with people that you trust. Discuss outline answers to past questions and swap ideas. Explain topics to each other and compare answers to the same questions. How does your approach vary from that of your friends? If you can work in this way, you will find that you become good at assessing the quality of each others' work. It is also not an uncommon experience that students are often more critical in their judgement of one another than teachers are of their students.

And finally....

1. Do not make worry a substitute for work.
2. Try to stay healthy – look after yourself.
3. Don't be complacent if you have moderate-to-good marks from the in-course work. You are in a good position, but a thorough understanding of the course will be important for future modules.

Ten tips for avoiding pressure during revision

1. Avoid stressful situations. Stress is contagious so don't moan to your friends about how much work you haven't done, or listen to people who tell you that they cannot sleep or that they are on tranquillizers.

2. Remember that you have a limited concentration span. Take short breaks every hour or so. Treat yourself. A short walk will refresh you as you realize that there is a world beyond the exam room.

3. When revising, test yourself to see how much you really do remember.

4. Don't drink lots of coffee, strong tea or energy drinks. Caffeine encourages adrenaline production and this induces stress. Tea has a lot of caffeine in it. Why not try herbal teas? (OK, because they can taste horrible.)

5. Don't turn into a recluse. The odd bit of socializing is good for you, providing you don't let it get out of control (see points 1 and 2).

6. Make sure you get enough sleep. Early nights are a good idea. Avoid burnout.

7. Don't leave all your revision to the last minute. Plan your work to build up steadily to the examination.

8. Don't make worry a substitute for work.

9. Try putting a few drops of perfume or aftershave on your wrists whilst revising. Do the same when you go in for your exam. Smell is a powerful memory trigger. It may work for you, too.

10. Stressed out? See your tutors. They are there to help you.

7.2 Examinations

7.2.1 How to fail your exams

Method one: do not write anything.

Method two: write everything you know about the topic mentioned in the exam question rather than selecting the appropriate information required by the exam question about that topic.

Method three: answer one question and get 100% for it and then ignore everything else. Simple arithmetic shows that if you have to answer three questions, then 100% + 0% + 0% = 33% overall mark. This is an automatic fail if 40% is the pass mark.

Method four: get out of your exam seat; go across to your best mate's place. Steal their paper and put your name on it. *But seriously...*

7.2.2 Styles of exam

Multiple choice questions (MCQs)

They are commonly used because they can test a wide range of knowledge quickly and easily. In addition, they can be marked rapidly and objectively. They are efficient at testing factual knowledge. Skilful question-setters can adapt the MCQ format to test some aspects of problem-solving skills. If you use practice MCQ banks in revision, make sure that you understand why options are correct or incorrect, and appreciate how the question relates to the topic as a whole. **Do not try to remember hundreds of dissociated facts.** If you adopt a surface learning approach to MCQ exams, you are doomed to failure.

A variety of formats can be used for MCQ tests. In the simplest there is a root followed by a number of branches, only one of which is correct. For example:

Frogs are:

    A. always green (F)

    B. amphibians (T)

    C. warm blooded (F)

    D. feathered (F)

    E. hermaphrodite (F)

A variation of this is where only one branch is *in*correct.

Other question types, sometimes referred to as multiple response questions or MRQs, may adopt a flexible pattern in which students are required to select the correct responses to a particular root question, where there may be a number of possible options given and *at least one response* will be correct. For example:

Hormones:

    A. are structural molecules (F)

    B. are energy storage molecules (F)

    C. may be peptides (T)

    D. may be synthesized from cholesterol (T)

    E. are involved in endocrine signalling (T)

With certain types of MRQs, all responses may be correct. MRQs have a number of advantages for assessment at university, but there is also a significant problem: to score full marks, all a candidate needs to do is to mark every option as correct. To get around this, MRQ tests are subject to negative marking where, for each incorrect answer, the candidate has up to one full mark (or a fraction of a mark) deducted. The rationale for negative marking is to deter guesswork.

For the sake of brevity, in the text below, "MCQ" refers both to the strict multiple-choice format and to multiple-response questions.

Strategies for passing MCQ tests: you should read through the paper carefully and note all the responses you know to be false. These may often, but not always, contain words like *"which of the following never..."* and *"which of the following always..."*. Mark as correct those responses you know to be true. Now read through the paper again. This time go through each question very carefully. Remember that every word in a question will have been chosen with care, and that you can often get clues about what the examiner is trying to test from the nature of the root statement and the possible responses. On the second run through, weigh carefully the responses. If you are still uncertain about the truth of a response, on balance it is probably better to ignore it.

Written examinations

Be well prepared and well rested for your exam. Last minute revision rarely pays off; most of the time it induces panic. If you have planned your studies and your revision using the advice in this book, you *will* have the knowledge to pass.

Try to take some time to relax the day before an examination and do not go to bed late. Aim to arrive at the examination hall in plenty of time, and find someone to talk to about topics other than the exam. Think positively. You will be nervous but so will everyone else. Continue to think positively. Take a few deep breaths and concentrate on relaxing tense muscles. A *mild*, controlled anxiety will enhance your performance and not detract from it. The trick is not to let the panic set in.

So, you are properly prepared, and you know you will pass. There are only three things that will stop you from passing now. These are:

- not reading the instructions
- not answering the questions
- not keeping to time

Read the examination instructions carefully. Make sure you understand the instructions and underline the key words. "*Answer <u>three</u> questions, <u>one</u> from <u>each</u> section*" means just that: answer **three** questions, answering **one question** in **each** of **three sections**. It does not mean "*answer all the questions on the paper*" nor "*answer any three questions*" nor "*answer the questions that you fancy answering*".

Answer the questions properly. Read through all the questions on the paper and turn over the paper to check whether there are any on the back. Carefully decide which questions you will tackle if you have a choice. If you do not have a choice of questions, start with the one that you feel most comfortable answering. Read through the question again, this time noting key phrases:

<u>Discuss</u>...
<u>Compare</u> and <u>contrast</u>...
<u>List</u>...
Write <u>short notes</u> on <u>three</u> of the following...

"*Discuss*" does not mean "*list*", nor does "*write short notes on...*" mean "*write a four page essay on...*"! Write an outline plan of your answer, as you did in your revision. It is a good idea to make your plan on the inside cover or first page of your answer book so that the examiners can see how you intend to tackle the question at a glance. If you do this, however, you should indicate clearly where the proper answer starts, particularly in answers for which you are invited to provide brief notes, which may get confused with your essay plan. For a half-hour question, your outline plan will take up five minutes or so, but it is time well spent. You are now ready to write a structured answer as discussed in *Section 5.5*. Make sure that your writing remains legible. Remember that your examiners need to read your words of wisdom.

Highlighted above are a few phrases that introduce questions and that cause students particular problems. There are many others. Table 7.1 lists some of the other common terms, together with a definition from the *Oxford English Dictionary* online, and additional notes if appropriate. You don't need to learn these by heart – just be aware of the various meanings when tackling questions in which they occur.

Deductive questions

Deductive questions ask you to use your knowledge and problem-solving abilities to interpret and evaluate new information, which may be qualitative or quantitative. This type of question can be scary because it is less easy to understand quickly what sort of knowledge is needed and how it should be manipulated to reach the answer than in factual essay-type questions. Try not to be put off by this, however, because deductive questions usually have several parts that structure your thinking and guide you to what is required. Also, remember that these questions are designed to test the skills and knowledge that you have acquired during your course rather than anything that is far beyond its scope. To tackle deductive questions successfully you should read the given information carefully. Then, you should work out what the information means, i.e. interpret it, evaluate it and reach conclusions, or propose a hypothesis, according to what the question requires. A cool head and confidence are required to tackle deductive questions and these can be acquired with practice.

Table 7.1 Analysis of common key terms used in examination questions

Term	*OED* definition	Notes
Analyse	To take to pieces; to separate, distinguish, or ascertain the elements of anything complex, as a material collection, chemical compound, light, sound, a miscellaneous list, account or statement, a sentence, phrase, word, conception, feeling, action, process, etc.	Examine in very close detail. Ask what the main features of the object are. Try to identify the principal points.
Assess	To evaluate (a person or thing); to estimate (the quality, value, or extent of), to gauge or judge.	Estimate the worth of the subject. Ask what the positive and negative points are. Consider whether views held by others on the subject are authoritative or not.
Comment on	To write explanatory or critical notes (*to*) *on*, or *upon* a text.	Identify the main topics of the subject and present your views based on your research to date.
Compare *	To mark or point out the similarities and differences of (two or more things); to bring or place together (actually or mentally) for the purpose of noting the similarities and differences.	To compare things you should show how two or more subjects are similar to one another.
Contrast *	To set in opposition (two objects of like nature, or one *with*, rarely *to*, another) in order to show strikingly their different qualities or characteristics, and compare their superiorities or defects.	To contrast two or more subjects you should point out the features in which the subjects differ from each other. You should indicate the criteria on which you base your judgement regarding these differences.
Define	To state precisely or determinately; to specify.	To define something you need to describe what you consider to be the exact meaning of the subject.
Describe	To set forth in words, written or spoken, by reference to qualities, recognizable features, or characteristic marks; to give a detailed or graphic account of.	To describe a subject you should set out the main features of that subject.
Discuss	To investigate or examine by argument; to sift the considerations for and against; to debate.	In discussion, you should set out the arguments for and against a proposition and draw your conclusions, indicating how these have been reached.
Distinguish	To make or draw a distinction; to perceive or note the difference between things; to exercise discernment; to discriminate.	When you are asked to distinguish between subjects you should highlight those features that make them different.

Term	*OED* definition	Notes
Enumerate	To count, ascertain the number of; more usually, to mention (a number of things or persons) separately, as if for the purpose of counting; to specify as in a list or catalogue.	When enumerating subjects you should list them, often including a brief description of individual items.
Evaluate	*a.* Math. To work out the 'value' of (a quantitative expression); to find a numerical expression for (any quantitative fact or relation). *b.* gen. To 'reckon up', ascertain the amount of; to express in terms of something already known.	When asked to evaluate a subject you should assess and describe its worth. In evaluating propositions, there may be evidence in support and evidence that refutes the idea. In a good evaluation, you should include both types of evidence and make conclusions, indicating how these were reached.
Examine	To inquire or search into, investigate (a question or subject); to consider or discuss critically; to try the truth or falsehood of (a proposition, statement, *etc.*).	When examining a subject, you should consider it in detail. As part of this process, you may also wish to evaluate the subject critically.
Explain	To assign a meaning to, state the meaning or import of; to interpret.	When explaining something you need to demonstrate why the subject is the way that it is.
Illustrate	*a.* To shed light upon, light up, illumine. *b.* To throw the light of intelligence upon; to make clear, elucidate, clear up, explain.	When you are asked to illustrate a subject, you should make it clear and explicit. It helps if you can support your illustration with evidence as to why the subject is as it is.
Justify	To make good (an argument, statement, or opinion); to confirm or support by attestation or evidence; to corroborate, prove, verify.	If you are asked to justify an argument or proposition, you should provide the evidence in support of that idea, showing how the conclusions regarding the concept were reached. For contentious ideas, you may need to present evidence that is in contradiction to your argument.
Outline	To describe the broad outlines or main features of; to sketch in general terms, summarize.	In providing an outline, you should only describe the principal points of the subject, preferably as briefly as possible.
Review	*a.* To survey; to take a survey of. *b.* To write an appreciation or criticism of (a new literary work, a musical or dramatic performance, *etc.*)	In reviewing a subject you should describe its principal features, exploring the reasons why these are considered important.
Summarize	To make (or constitute) a summary of; to sum up; to state briefly or succinctly.	In summarizing a subject, you should draw out its principal points, omitting details and examples. Summaries are brief forms of outlines.

*These are often combined. Sometimes you may be asked to compare two topics without being asked to contrast them. If you feel the need to show how they differ, state this explicitly.

7.2.3 *Plan your time*

Check the length of the exam, estimate how long you have for each question, and stick to it! Three hours to answer four questions allows 40 minutes per question, with twenty minutes to be split between the beginning and the end of the exam. This time is to read the questions before you start, and to check your answers at the end. Allocate time proportionally. The first 20% of marks are very easy to earn. The final 20% are the hardest marks to pick up. This is why you must attempt the required number of questions rather than relying on doing just two or three very well.

> **Do not be tempted to spend a long time on your favourite topic at the expense of other questions.**

You are better off scoring 60%, 56% and 58%, giving an overall mark of 58% than you are scoring 80%, 60% and 20%, giving you an overall mark of 53.33%, just because you ran out of time on the last question.

7.3 What is looked for in an examination answer

Essays written under examination conditions cannot be as finely crafted as essays written in your own time at home, but there are many similarities in how they are assessed. The section below gives an idea of how different qualities of answer are graded.

First class answer

This displays a complete understanding of the question. Nearly all of the information given in lectures, tutorials, etc. will be included, but there will also be evidence that you have read around the subject. Answers will be constructed from a variety of sources, logically organized, well presented, and substantially free from errors. If appropriate, the first class answer demonstrates originality of thought or approach and will also display insight.

Upper second class answer

This is slightly less outstanding, but will display a sound understanding of the question, containing all, or nearly all of the information given in lectures as well as some evidence of reading around the subject, although not to the extent expected of first class answers. The answer will be largely error-free and will certainly not have any major errors of fact. It will be logically presented and the better answers in this class will demonstrate your ability to manipulate information.

Lower second class answer

This is generally sound, repeating the material presented in lectures. It will probably not show evidence that you have read around the topic, but it will be free from major errors. The presentation will usually be less logical than answers deserving of higher classes.

Third class answer

This provides evidence that you have some understanding of the question. Such answers are typically incomplete and may betray a poor appreciation of the subject.

Important points will not have been addressed and a third class answer may contain a quantity of irrelevant material. The presentation of such answers is generally poor.

A pass mark or below

This answer must be a serious attempt, containing some relevant information to show that you have a limited understanding of the question. It will contain serious omissions or errors. If your answer is inadequate, lacking in substance or understanding then you will get a fail mark.

7.3.1 In-course assessments

A significant proportion of course marks may be gained through some form of in-course assessment rather than from examinations. The tasks that you will be set will vary from course to course, but you should be clear what proportion of the overall mark will be derived from in-course assessment, and how these marks can be earned. The important thing with in-course assessment is that you must **always hand something in,** no matter how badly you think you have done. You will never get less than zero for any piece of work attempted, and zero is all that you will get if you fail to hand work in, or if you do not turn up for a test. It is polite and also sensible to inform the person(s) running your courses if you are ill; they are more likely to be sympathetic and help you make up the work you missed, thereby allowing you to gain coursework marks you might otherwise have lost, than if you do nothing.

7.3.2 Summary of advice on examinations

If you want to succeed in your exams and do the best that you can, our top ten tips are given below.

Checklist for examination success

- Scan the paper to give yourself an idea of the scope and range of questions available.
- Make an initial selection of the questions you think you can do best. Don't worry if you don't think you can do the requisite number at this stage.
- Start with the question you think you can answer best.
- READ THE QUESTION PROPERLY.
- Draw up an answer plan (on your answer book cover is a good place). Ensure that you remain focused on the required information.
- Calculate how long you have per question/section and stick to the allotted time.
- Do answer the requisite number of questions even if you feel less confident of your last choice.
- If your mind goes blank then think about everything that you might know about a topic.
- If all else fails then try another question.
- *Remember*: the first 20% of marks are much easier to gain than the last 20% in any question.

Good luck in your exams!

If you have read this section, it probably means that your exams are close. If so, let us hope that this is not the first time that you have opened this book!

Chapter 8 Taking a year out

Some universities operate schemes that give students the opportunity to broaden their skills and experience by taking a year out of full-time study. This is usually, but not always, offered at the end of your second year. If available, this contributes in a positive manner to your degree. You may be able to undertake, for example, a placement in a relevant industry, a year studying at a foreign university, or voluntary work.

When you are making your decision about which university to attend, you should check which universities offer such schemes if you think that you would like to take a year out during your degree. Make sure that you also know what criteria are required to qualify for the scheme, such as the minimum level of academic achievement required to enrol on the programme, the relevance of any modules you are taking, and a timescale indicating what you must do in order to make your application. For example, if you are taking a year abroad in a country where English is not the first language, you will need to ensure that your language qualifications are suitable for the location in which you intend to study.

Departments where such schemes are well established are likely to have appropriate systems for helping students find a placement, and provide support while their students are away. They will also have tried-and-tested processes for coordinating with, and monitoring, the host organization. If the department does not normally run such schemes, you will need to gain appropriate permission and will also need to be prepared to do more of the basic work yourself to find suitable placement organizations and follow the relevant application procedures.

Some universities take account of your performance during the year out when calculating your final degree result, so you should check the assessment weighting and how this is applied. Other institutions regard the year out simply as a pass-or-fail year and, depending on your success, will indicate that you have been successful by awarding you a degree with a title that recognizes your achievement.

Work that can be done during the vacations does not normally require permission from your university but can be equally valuable in providing useful experiences and helping you to develop relevant skills.

8.1 Why bother? The pros and cons of taking a year out

So, why bother prolonging the time it will take for you to gain your degree? Daunting though a year out may seem at the start of your undergraduate journey, there are many reasons why you should consider the matter seriously.

Both an industrial placement and a study year abroad will broaden your experience of life by presenting new and interesting challenges. The activities you undertake during this additional year may also give you a firmer idea of what sort of career you might prefer and where in the world you might like to work. While on these schemes, you will meet and work with new groups of people, experience a different environment, work to different timetables and, if you choose to study abroad, you will encounter a different culture, and perhaps be required to communicate in a different language. In the global economy, this will be a valuable asset in your search for future employment.

On a personal level, you will develop a greater degree of self-reliance, become more efficient at managing your activities and time, become a better communicator through your interaction with a variety of different people, and will be able to use your initiative to deal with different sets of circumstances and problems. These skills will be recognized by any future employers and are likely to give you an advantage when applying for jobs. You might also make useful contacts to whom you might apply for a job later in life.

There are direct benefits of the year out in terms of your studies and future employment. First, most students come back to university more confident and more self-assured – as well as one year older of course – and this additional maturity usually stands them in good stead when addressing their final year studies. They often end up with very good degrees. Then, of course, the year out experience is valuable additional material for your *curriculum vitae* (CV) when applying for that first job, and will typically place you at a distinct advantage in the graduate jobs market.

Taking a year out means that some of the peer group with whom you started your course will have left by the time you return to your academic studies. In reality, some of your peer group will probably also have taken a year out and will return to university studies when you do. So when you start your final year you will have a nice mix of old and new friends.

Finally, having a year out may have a major impact on your financial situation because, in addition to living expenses and accommodation, you will still be required to pay a proportion of the university fees, to cover the costs of supporting and administering your year out. However, many industrial placements offer a salary that will cover living expenses and accommodation throughout the year, and so some students may even manage to pay off some long-standing debts! A year at a university abroad or of voluntary work, however, may leave you seriously out of pocket, and you should investigate the possibility of obtaining a bursary or sponsorship as a means of supporting yourself in these circumstances.

8.2 An industrial year out

When applying for an industrial placement, you will be in competition with students from around the country, as would happen if you were applying for a normal job. It is vital that you complete an appropriate application form and submit a suitable up-to-date CV. Your departmental tutor will need to write you a reference indicating your suitability for a placement, which may also require approval by the university. If you are invited for interview, you must give the best impression you can to your interviewers. If you are then offered a placement, it is your responsibility to make the most of the opportunity.

A major benefit of working in a research setting is the opportunity to gain a range of subject-specific technical skills and to use the latest research techniques and equipment. Another benefit is that you will be conducting meaningful research that may even lead to a research publication that can be listed on your CV. You will also be able to experience the group dynamics and teamwork required for successful research environments. For many students, the experience of an industrial year out is also very helpful in helping them decide whether they would like to move into research after completing their degree – at least to PhD level and possibly even moving on to an industrial career thereafter. At the end of the placement, most students write a report and may produce a presentation such as a poster or a mini-seminar, so good communication and presentation skills will be important here.

You may be uncertain about whether you will have lost the habit of academic working during a year out and how much academic material you have forgotten, and may believe that integrating back into your university course will be difficult. This is (almost) never a problem. Much more common is to find that the extra year's maturity helps students to focus more effectively on their end goal – to get the best possible degree. You may actually find that the relevance of the academic material becomes clearer having done a year of 'real-life' science. The following sections offer suggestions on how to derive most benefit from the year in an industrial placement.

8.2.1 *How to make a success of your industrial placement*

Successful placements require a very high level of commitment and you should work hard and behave in a professional manner at all times. This means that you should show an interest in the project you are given, be enthusiastic about the tasks involved, and be prepared to put in appropriate hours to get a job done. It is important that you ask questions to make sure you understand what is required of you and that you can show some initiative and work independently. The more expertise and reliability you can show, the more your project leader is likely to trust you to take on more advanced and, quite likely, more interesting work using a greater variety of techniques. This will help you to develop your practical skills further. Make the most of working as part of a group by observing and learning how the other team members deal with issues that arise. You will meet a variety of people and have an opportunity to talk with many about their roles and what career paths led them to that position. This may give you ideas for the future, but also, importantly, you may decide to ask one of them for a future reference.

Help yourself settle in quickly by making the most of the social opportunities on offer; there may be a sports centre and organized activities that you can join. If there are other students on industrial placements at the same company, see if you can share a flat together – you are all in the same boat and can share your experiences.

Many universities ensure that your industrial supervisor is given clear guidelines on the project type and supervision that needs to be provided and there will often be an Industrial Work Profile to be completed by the student and industrial supervisor during the course of the year. This will describe skills acquired, work done, goals achieved, etc. Typically you will remain in contact by email and telephone with your university during your year out and an academic tutor from your university may well visit the company site to meet with you and your supervisor and check that all is well.

Problems may arise if your expectations do not match with the position you are given within the placement. It is possible that the company is assessing your abilities before moving you on to working in a more responsible situation, but if things don't improve within a few weeks it is important that you talk to your industrial supervisor and your University tutor so you can sort it out. If you are not enjoying the work, again, you must find someone to talk to about this; you may have unrealistic expectations of the work, or you might be allowed to swap the focus of the work to something you find more enjoyable.

In the event of a personality clash with your supervisor, you might need to call upon all your energies to maintain a professional relationship. Again, however, it is worthwhile talking to your university tutor and to others working in the team; they may be able to give moral support and advice especially if they have a similar relationship issue with the supervisor. If difficulties persist, you should approach the Human Resources section of your place of work as it is their job to address such issues.

Finally, do **not** expect to have a successful industrial placement if you:

- spend a lot of time socializing
- make a lot of mistakes
- waste time
- turn up late for work
- appear uninterested
- fail to show initiative
- ignore the other group members

This will create a very bad impression and may endanger the chances of any other student from your university obtaining a placement there in the future.

Checklist for a successful industrial placement

- Be committed to the work – be prepared to work the required hours to get the job done.
- Act professionally.
- Show enthusiasm.
- Ask questions if you do not understand anything.
- Develop your practical skills.
- Complete the industrial work profile as you progress through the year.
- Learn from others in the team.
- Join in with social activities.

8.3 Study year abroad

You may be attracted to a study year abroad because you are interested in discovering more about a particular culture and/or you would like to improve your language ability whilst continuing your academic studies. Opportunities exist for students to visit a variety of places throughout the world such as the USA, the Far East, or even Australia. The Erasmus scheme applies to students studying at European universities. Taking a study year abroad involves spending a year in a foreign university, attending courses and, most likely, conducting a project in an established research group. To ensure parity between the academic standards in your home university and the institution you visit, your home university will already have checked that a suitable range of courses at the correct level is available for visiting students. In addition to modules in your subject area, there are likely to be modules covering a very different range of topics from those offered in the UK; for example, in Singapore, you may be offered Islam in Contemporary Malay Society, and in the USA modules covering the History of Black African Women in Society have been offered.

To qualify for the scheme you will be required to make an application and to obtain a reference from one of your tutors. Universities often have well-established links with particular institutions abroad, so there is less likely to be competition for those places from students from other universities, than when students are competing for industrial placements.

By opting for a study year abroad you will be continuing your academic studies in a different culture and also possibly in a different language. Integrating into a student group with a very different culture can be daunting, but if successful, will bring great rewards, as illustrated by the comments from a student who spent a year in Singapore:

> "I was very lucky to enjoy the support and friendship of a close-knit lab who introduced me to the Singaporean life, especially the wonderful array of food! I was invited to take part in a traditional Chinese wedding ceremony, which was a highlight of my time there. Other foreign students came from numerous other countries and I now have friends in Canada, America, Europe, China, Australia and many more. ....It was an amazing opportunity which I would repeat in a flash."

8.3.1 How to make a success of your study year abroad

In many ways, the recommendations for a successful industrial placement also apply to a study year abroad. It is important to be committed, that you attend all the teaching sessions, and spend an appropriate time working on your project. You will be interacting with a research group in a similar manner to students on the industrial placement, so it is very important to give a good impression especially if you might want to return to work there in the future. Although it may be difficult to break into friendship groups that are already established, especially if there is a language barrier, there are bound to be societies and clubs through which you can establish a circle of friends. If you have learned much more about the culture of the country you visit and have made new friends by the time you return to the UK, you will have fulfilled a large part of the aims of a study year abroad scheme.

It could be very easy to feel isolated in a foreign country, especially if you find it hard to get to know other students. If this feeling continues for more than a few weeks, it is VERY important to talk to somebody. Your project supervisor or someone else at the University who is involved in organizing the visits is a good place to start. You must also let your tutor at home know (by email) how you are feeling because they can offer advice and contact relevant people at the institution where you are. The VERY WORST THING you can do is to bottle up your feelings and not tell anyone. If you discover one of the modules you have chosen is not what you expected, you should ask if it is possible to change to a different one – and keep your tutor at home informed of what you are doing. For a very few students, problems such as this can lead to depression, poor attendance, and possibly a failure to complete the year. Remember the advice we have already given on asking for help.

Checklist for a successful year abroad

- Attend all the required teaching sessions.
- Show enthusiasm for your project.
- Ask questions if you do not understand something.
- Show initiative and independence where appropriate.
- Be prepared to work the required hours to get your work completed.
- Be aware that although you may feel isolated at first, this feeling is likely to pass soon.
- If a feeling of isolation continues TALK TO SOMEONE ABOUT IT.
- Make new friends by joining societies and clubs.
- Explore the country and the culture.

Chapter 9 Your final year

If you have followed the advice in this book, eventually, and this will happen sooner than you may think possible, you will enter your final year. After that, you will have to face the big bad world. To make sure that you are properly prepared, you should plan to make the most of the opportunities open to you. One of the key features of your final year is your project. You will also need to prepare for the world beyond the ivory towers of Academe. This section provides advice on optimizing your performance in your final level of study.

9.1 Final year projects

In your final year, you will probably carry out a research project. This involves considerable effort each week. Like practicals, projects may or may not be laboratory-based. A project that is not laboratory-centred may involve, for example, the preparation and presentation of a dissertation, exploring important topical questions. This requires assembly and critical reading of previously published research papers. As such, much of this type of project may be library-based. Another popular format for such so-called "dry" projects is to undertake a survey. Some students prepare education and information packs and even film storyboards, communicating scientific ideas to a lay audience; others plan applications for research grants. Alternatively, projects may entail the development of a particular type of computer software. The only limit on this type of project is your imagination.

Laboratory projects involve tinkling the glassware (these days, however, glass has largely been replaced with plastic!). They demand that students undertake individual practical research projects under supervision. Such projects will typically be carried out in the research laboratory run by the project supervisor, instead of in large teaching laboratories. This gives students the opportunity to work alongside research students and technicians, from whom they can learn a great deal.

As in all other areas of your undergraduate work, for success in your final year project you will need to be strategic, to plan, to be diligent in following your plan, and to present your project clearly. Regardless of the type of project, you must first identify the aims of the project: what is the problem and what are you trying to achieve? From this you must identify the tasks that need to be done to achieve the aims, including the objectives and any methods or approaches that it will be useful to follow – the published literature will help in this. Set yourself a feasible timescale but also be flexible to allow for unexpected avenues of research. Your results must be clearly presented, with your conclusions related to both the project aims and the current accepted understanding of the project topic in the literature. Finally, the whole must be succinctly presented in a report that conforms to the subject's academic style. Ensure that you do not become 'sloppy' as this will affect the final product.

Whatever type of project you undertake, you will be acquiring many skills that will be of great benefit in your future career. These include planning and organizing your time, developing a critical approach to published material, and learning how to present your findings in the most appropriate manner.

9.2 Planning for life after university: thinking about careers

Sooner or later, the day will come when you will venture out into the world of work and get a job. Some of you will be glad of the opportunity to earn a salary to start to pay off your student loan and to support yourself as soon as you have finished your degree; others, no doubt, will delay this event and choose to continue in higher education, aspiring to a higher degree. For both groups, success will come to those who have clear plans, and the strategy for securing a good job is the same as the strategy that is most likely to secure a top PhD position. There are numerous sources of help, support and advice but, ultimately, the only person who can make appropriate decisions regarding your future is you. The more carefully you plan for your future, the more likely you are to succeed. This section is intended to give advice on a strategy to adopt.

9.2.1 Gather information

Some students have a very clear desire to follow a particular career path after their first degree; others have little idea what they want to do. Whichever group you belong to, the most important first step to securing a successful career is to get as much information as possible about what is available to you. This includes informing yourself of several alternatives so that, should things not work out for you, you have something to fall back on.

9.2.2 Assess your skills and qualifications

Employers often draw up lists of essential attributes and desired attributes when seeking to employ graduates, so if you don't have any of the essential attributes, even just one, you will be wasting your time putting in an application. Therefore, the next step in successful career planning is to assess whether you have the necessary skills and aptitudes for the jobs that interest you; it is very important to take a critical look at your abilities, strengths and weaknesses. It may be easy to identify abilities and strengths but it can be hard looking at your weaknesses. Unless you are honest with yourself and recognize your limitations, you will apply for positions that are unsuitable for you and you will end up with a long string of rejections, which is very demoralizing. You should also focus on your interests and motivations. A well-paid job that bores you rigid may be tolerable for a few weeks or months, but it is no basis on which to build a successful career.

9.2.3 Make a short list of available jobs

Having decided on what interests you, and whether you have the necessary attributes and qualifications, the next step is to find out what is available. Your university Careers Service will undoubtedly provide a rich source of information, and recruitment sites on the internet are good places to find job adverts. It is also well worth looking at the jobs sections in newspapers and in scientific journals such as *Nature* and *New Scientist*. The range of jobs for science graduates is surprising so, hopefully, you should find some that appeal to you. While you wait to spot the job of your dreams, bear in mind that these may take a little time to materialize. While it is very likely that your degree will ultimately provide you with an entry into a rewarding career, be prepared to take low-paid, undemanding work until the right job comes along.

9.2.4 Make the most of first contacts

Sometimes, employers give contact details for applicants to discuss posts before making a formal application. This is always worthwhile; if nothing else, it shows that you have an interest in the position being advertised. Even though such contact is generally informal, it is an opportunity to sell yourself and thorough preparation will pay dividends. If there is a telephone number to contact, do your homework: find out as much as you can about your potential employer and use this information to frame a list of pre-prepared questions. It is essential that you understand the nature of the job for which you apply; failure to do so could leave you in a job in which you soon become very unhappy. If applying for a PhD position, read the publications from the lab to which you want to apply. Remember that it is probably **not** a good idea to start with questions focusing on the social side of the job! If the contact involves a visit, make sure you dress appropriately; managers of field centres rarely wear high heels to work.

9.2.5 Your curriculum vitae

While the prospect of finding a job may seem a long way off when you are new to university, planning can never start too soon. In parallel with your PDP, it is a good idea to draft your CV at a very early opportunity. If you plan to take a year in industry, you will need to prepare a CV very soon into your undergraduate career, but even if this does not appeal to you, the earlier you draft a CV, the better. Having prepared your initial electronic document it becomes a relatively trivial task to keep it updated with significant milestones during your academic career. The earlier you start, the more time you will have to produce a CV of which you can be proud.

A CV is a document listing your personal details, and includes a history of your education, your qualifications and achievements, and other information such as your interests. Your prospective employer(s) will invariably want to see one in order to assess how suitable you will be for their job. Your PDP will help you to remember the important events that occurred to you at university, and will also be invaluable if you are asked questions about how you developed your skills and attributes throughout the course and what skills you might bring to the job.

Creating a CV

Since employers will gain the first impression of you through your CV, it is **vital** that you present it in the best possible format; a poorly presented CV indicates that you are either unable or cannot be bothered to make an effort, and your application is likely to be directed straight into the waste paper basket! Remember, also, to tailor your CV to fit each job for which you apply since each one is likely to request slightly different attributes and skills. Things to remember are:

- keep the information well spaced with important details clearly visible
- include information that is relevant to the job and do not include irrelevant information
- keep your language simple and to the point
- avoid any unexplained time gaps in your education or employment; it is better to be up-front about gaps in a CV than to ignore them or, worse, try to cover them up
- ALWAYS check the grammar and spelling (use a dictionary)
- ask a friend to look over and comment on your drafts
- expect to have to redraft the CV several times before you achieve the final result
- use a simple font such as Arial or Times New Roman, rather than something more casual such as Comic Sans or fancy such as *Monotype Corsiva*
- use good quality paper (white) and a high-resolution printer

What to include in your CV

- Your personal information: your name, address, contact telephone numbers and email address. Remember to make sure your email address does not give the wrong impression: as before, 'studmuffin' or 'sexy_babe' are not appropriate!
- Previous education: in chronological order starting with the most recent.
- Any work experience: list everything in chronological order from the most recent – make sure the dates you give are precise.
- Other skills and qualifications: such as a driving licence, first aid training, special ability in languages, musical and sports qualifications, or leadership and organizational skills that might be highlighted by particular activities in which you have engaged.
- Your interests and activities: anything that shows you have more qualities than can be ascertained from your education and previous employment.
- Referees and their contact details: it is usual to be asked for two; ideally one of your tutors and, if possible, one from an employer, although if you have not had work experience, two different academic referees can be listed. Do remember to get permission from them before listing them on your CV.

9.2.6 Making an application

Once you have got this far, you are ready to make your application. Take care with your presentation and answer all the questions on your application honestly. If you are caught not telling the truth, the consequences could blight the rest of your career. Indeed, it has been suggested that you could lay yourself open to prosecution for fraud. If asked to supply a CV with your application, make sure it is well presented: be careful in your choice of fonts, colours and paper. Stencil script in pale green on lilac paper will certainly stand out from the crowd, but will it really convey a message that most employers want to receive? Likewise, take care over any covering letter that you use to accompany your application. This should be short: ideally no more than one side of A4 paper. It gives you the opportunity to explain, succinctly, why you feel that you are particularly suited to the job and also allows you to explain any potential problem areas on your CV, like the poor marks you got for certain modules. If you follow this advice, then invitations to interviews will hopefully start arriving. Again, preparation is the key and if you have prepared properly, then the job offers will soon follow.

Future employers and PhD supervisors will usually require two references about you so your CV should contain the name and contact details of two referees. If you are applying for employment, one referee should be able to comment on your recent academic studies and the other should ideally have some experience of your performance in the workplace, to provide comments about your employability skills and attributes. A good choice of academic referee would be someone who knows you well such as your personal tutor or your project supervisor. As mentioned above, it is polite to contact your chosen referees to ask if they are willing to act in this way for you. To write an effective reference for a specific post they need to know the specifications of the job you are applying for. Try to make the work easy for them by sending these details rather than simply referring them to a website. It is also useful to let them see your proposed letter of application or personal statement to help them to focus their reference. They may also be able to make useful suggestions on content. Ensure that you and your referees are clear about how the reference is to be communicated; some PhD application processes require you to enclose written references with your application, others ask the referee to contact them directly and some companies may expect email or telephone references.

9.2.7 The 'milk round'

Graduates are sought after by employers. By the time you graduate, you will have acquired and practised numerous 'transferable skills' that make you useful to employers in the sciences and beyond. In recognition of this, large companies organize themselves to visit selected campuses for the purpose of recruitment in events known as the 'milk round'. If your university participates in this scheme you will have the opportunity to discover more about the companies in which you are interested and may even be offered on-the-spot interviews. If you are unsure about what career you might like, this is an ideal opportunity to gather information from all the employers attending by talking to the staff manning the stands. You might then be able to refine your ideas.

9.3 Be prepared for opportunities when they arise

Whatever career you decide to pursue, the take-home message from this section is that success will come with good preparation. Find out as much as you can about yourself, your aspirations and motivations and do the same about your potential employers. A final piece of advice is to keep an open mind. Circumstances sometimes close off options but this should be seen as an opportunity rather than a threat. If you find that for any reason you cannot follow the career path that you had planned, be prepared to explore alternatives. These can be very rewarding if approached with a positive attitude.

Appendix:

Working with *Word*, Perfecting *PowerPoint*, Excelling in *Excel* and Becoming Adept at *Access*

Long experience has shown that many students are unaware of how much the *Microsoft Office* suite can offer to help with their studies. It is assumed that students will have learned how to make the most of these powerful tools at school but sadly, this assumption seems to be ill-founded. To help address this, we have provided a guide to some of the most powerful features of the principal *Office* programs so that you can make the most of them when analysing and presenting your work.

When writing essays and practical reports it is easy to get away with using *Word* as a fancy typewriter to produce reasonable results. However, *Word* is a very powerful program with features that will become invaluable when creating complex documents such as your Level 3 project report. Likewise *PowerPoint*: this is easy to use to create simple presentations but also has a number of sophisticated features that enable you to lift your presentation significantly. The *Excel* spreadsheet enables you to manipulate complex data with ease and to produce dynamic visual presentations that can be exported to other programs in the *Office* suite. Finally, when storing and analysing complex data sets, an *Access* database can be invaluable, particularly where different data sets can be linked. The sooner you start to explore the features of the *Office* suite, the more confident you will be and the better will be the standard of the work that you produce.

Because of the coding that underpins *Access*, it will only work in a Windows environment; there is no Mac equivalent. If you are required to use *Access* as part of your course, you should receive bespoke training on its use and, of course, your Faculty or University will ensure that you have access to the Windows version of *Access*.

The following can only be an overview of some of the more advanced facilities available in the *Office* suite. Since the software is regularly updated, any detailed instructions we might include here of how to actually achieve the desired outcome would quickly become outdated. Therefore, we would recommend that you refer to the dedicated 'Help' provided within each package since it will contain the latest information. There is also a vast amount of up-to-date information online that you can consult.

Perhaps two of the most important pieces of information you should always remember when producing electronic documents are:

- SAVE YOUR WORK REGULARLY, in case you have a disaster with your computer that will lose your latest version of your work;
- KEEP BACKUPS of all your important work.

Faster ways of working

The Quick Access toolbar

Using the Windows program, you can create your own individualized toolbar containing the functions you most often use (for example, new, undo/redo, insert, save, print layout and print), regardless of which taskbar or ribbon they are usually found in, and thus you can avoid wasting the time clicking through different menus or toolbars to find them. If you use the Mac version of *Office*, you need to click the view menu, then toolbars> customize toolbars and menus.

The Shortcut menu

Right clicking the mouse when it is positioned in a part of the document you want to modify provides you with a shortcut menu, which allows you to 'shortcut' to some operations thereby avoiding the need to go through the Tabbed Menus. If using the Mac version of *Word*, as long as you have activated the right click function (also known as 'secondary click' in the system preferences menu), then you can use the shortcut menu as described above.

Shortcut key combinations

Shortcut key combinations are invaluable for increasing the efficiency with which you can work. Again, they allow you to achieve commonly used tasks, which may otherwise require clicking through several menus, by holding down two or more keys simultaneously. For the most common tasks, there is more than one shortcut available. There is an abundance of help available on the internet, with many pages and even video clips giving live demonstrations on how to achieve particular results. Be aware, however, that as with all internet resources, some are better than others and a few are positively misleading; pick your source with care. The table below indicates some of the most common shortcuts; the list is by no means exhaustive.

Table A1 Shortcut key combinations.

File-associated shortcuts	Windows key combination	Mac key combination
Office Button/New	Ctrl+N	cmd+N
Office Button/Open	Ctrl+O	cmd+O
Office Button/Close	Ctrl+W	cmd+W
Office Button/Save	Ctrl+S	cmd+S
Office Button/Print	Ctrl+P	cmd+P
Office Button/Exit	Alt+F4	cmd+Q
Editing shortcuts		
Undo	Ctrl+Z	cmd+Z
Redo	Ctrl+Y	cmd+Y
Cut	Ctrl+X*	cmd+X
Copy	Ctrl+C*	cmd+C
Paste	Ctrl+V*	cmd+V
Select All	Ctrl+A	cmd+A
Find	Ctrl+F	cmd+F
Replace	Ctrl+H	cmd+Shift+H
Formatting shortcuts		
Format/change case	Shift+F3	Shift+F3
Small capitals	Ctrl+Shift+K	cmd+Shift+K
Subscript	Ctrl+=	cmd+Shift+=
Superscript	Ctrl+Shift+=	cmd+Shift++
Apply list bullet	Ctrl+Shift+L	cmd+Shift+L
Make bold/not bold	Ctrl+Shift+B	cmd+B
Italicize/unitalicize	Ctrl+Shift+I	cmd+I
Underline/remove underline	Ctrl+Shift+U	cmd+U

* Shift+Delete also works in Windows versions to cut highlighted text, as does
Ctrl+Insert to copy text and Shift+Insert to paste material from the clipboard.

Performing bulk operations, rapidly

There are various options available when you want to perform bulk operations, as
listed below:

- To repeat an action several times in different places (for instance italicizing
  different sections), you can select all the words that need changing first and then
  perform the operation just once. In the Windows version, this is done by selecting
  the first section, then holding down the Ctrl key when selecting every subsequent
  item that needs the same change, before applying the required operation. If using
  the Mac version, you hold down the cmd button rather than the Ctrl key.

- The 'Find and Replace' facility is extremely useful but be very careful when using
  the 'Replace All' feature, since text for replacement may occur in contexts that you
  do not expect, for example as part of a longer word. Global replacements can end
  in some very strange and unwanted issues. For instance, replacing all occurrences
  of '*etc*' by '*and so on*' using the 'Replace All' button in a document containing the
  word '...stretch...' would result in 'str*and so on*h'.

- Macros are short computer programs you can create to perform repetitive tasks rapidly and accurately. They are ideal if you want to divide up a long nucleotide sequence into triplets separated by a space, or to change every instance of a binomial in normal font to italics. They are designated a shortcut key combination of your choosing so that you can set it running from the keyboard when you need it to perform the required operation.

- Moving a whole chunk of a document can be done very rapidly by dragging and dropping, which involves selecting the relevant section, then keeping the left mouse button depressed while you drag the section to a new position. You can also duplicate the chunk by depressing the Ctrl key as you drag the copy to its new position. While the Mac cmd button is frequently used in the same way as the Windows Ctrl key, when moving text in the Mac version of *Word* you need to hold down the Alt key. Some people find this all a bit cumbersome and simply either cut and paste or copy and paste the desired text. When creating *PowerPoint* slides freehand (see below), this is a useful technique for repositioning text boxes.

Overwriting

This is the process by which you can select a portion to be changed then write over it directly. This is faster than selecting the section and deleting it before writing in the new text.

Working effectively with *Word*

Although you can achieve a very presentable essay simply by exploiting the buttons on the ribbon that you see when opening *Word*, there are many more features in this software that should be explored, as explained in the following sections.

Navigating around large documents

At some point, you will need to produce a large report. While it is relatively simple to navigate around an essay of, say, 1000 words, this task gets much harder when creating a project report, which is ten times that size. You can use the 'Page Up' and 'Page Down' to scroll through your short document but what happens if you are working on a very long document and you need to refer to a section some way away from where you are currently working? Two useful options are listed below:

- splitting the screen into two frames, each of which may be navigated independently. You can now scroll to the section that you want to review in one frame while leaving the cursor at the point where you are currently editing in the other frame. The screen can be split horizontally or vertically depending on your preference.
- applying different 'styles' (such as heading types) to your document (see below) then using the Navigation Pane or the Outline View to display the document structure. Clicking on specific items in the structure displayed allows you to jump easily to the section that you require. If you want to reorganize sections of your document, you can do this in Outline View simply by dragging and dropping selected sections up and down the list.

Another useful navigation trick is to find particular texts. This requires you to be precise with the text for which you are searching, however. As well as finding words and spaces, you can use special codes to find paragraph marks, tab marks, page breaks, any unspecified character (this is particularly useful when combined with specified formats), non-breaking hyphens, non-breaking spaces and so on.

Formatting your document

Fonts

You can easily control how your text looks by changing the font that you use. There is a huge variety of fonts, both in style and in size, from which you can choose, although for some purposes, you may be constrained in the font(s) that you can use. Try not to use 'funky' fonts just to be different – mostly, they are annoying. Also, make sure that you pick a size that is legible to elderly professors.

For standard essays, Calibri (Body) font size 11 points or Arial font size 11 points are generally acceptable. **Bold**, <u>underlined</u> and *italicized* text is useful for emphasizing and highlighting particular words or sections.

Your text needs to be easy to read. Most fonts are 'proportional', with the letter 'i' taking much less width than does the letter 'm', or 'w'. While this is fine for most texts, in some cases, however, it is better to use a font in which all letters take up the same width, such as when presenting nucleotide sequences. As an illustration, the first ten codons of the bla_{TEM} gene from Tn*3* are presented first in the Courier New font and then below that in the Arial font. **Note how the sequences no longer align properly when using a proportional font.**

```
5' atg agt att caa cat ttc cgt gtc gcc ctt 3'

3' tac tca taa gtt gta aag gca cag cgg gaa 5'
```

5' atg agt att caa cat ttc cgt gtc gcc ctt 3'

3' tac tca taa gtt gta aag gca cag cgg gaa 5'

Non-breaking spaces and hyphens

If the page formatting would cause scientific binomials, such as *E. coli*, to be split over two lines, or numbers to become separated from their defining units, non-breaking spaces can be used to prevent this. To insert a non-breaking space, press 'Ctrl+Shift+spacebar' if using the Windows version of *Word*; for Mac users, the key combination is alt+shift+spacebar. In a similar way, to avoid splitting β-lactamase, or any other hyphenated words, across two lines, you can use a non-breaking hyphen. Both of these types of formatting can be inserted using specific key combinations. Ctrl+Shift+hyphen will insert a non-breaking hyphen when using the Windows version of *Word*; if you are using a Mac version, you will need to press cmd+shift+hyphen. You will be able to identify the presence of non-breaking spaces and hyphens as tiny circles between the relevant characters when the Show/Hide icon is used to visualize the formatting.

Special characters

As well as changing the appearance of text, *Word* has the facility to allow special characters to be inserted. Particularly useful in science are the Greek letters that are used in names and also to represent units and mathematical functions. With Greek letters, if you know the equivalent in 'normal' text, changing the font to 'Symbol' will change the letter to its Greek equivalent, as illustrated below:

- changing an 'a' or a 'b' to the symbol font allows you to write about α- and β-haemolysis, respectively;
- altering a 'c' to a symbol allows you to write about the χ^2 test;
- the prefix for 'micro' is a converted 'm', giving 'μ'.

Alternatively, you can use the 'Insert' tab and click on 'Symbol' to open a dialogue box that lets you pick from a very wide range of characters and letters. This is how you can add the umlaut to 'Müller Hinton agar' and add an icon to advertise your telephone number (☎: +44 [0]113 ....).

Paragraphs

The Paragraph dialogue box allows great flexibility in the appearance of paragraphs. You can choose the line spacing using pre-set values or define your own in order to fit the text into a specific space. Students are often asked to submit work with double spaces between lines, since it allows room for annotations when it is being marked. Depending on the font being used, single-spaced text can appear somewhat dense.

You can also decide how much space, if any, you would like before and after a paragraph – it is often better to set just ONE of these to keep things simple and avoid spacing problems when dealing with consecutive headings. You can decide how much, if at all, the first line of the paragraph should be indented. Hanging paragraphs, where every line *except* the first is indented, are useful when dealing with a series of itemized lists.

> Indentation on both sides of a whole paragraph can
> make it stand out from the main body of the text.

You can also set how the paragraph is aligned on the page, as illustrated below:

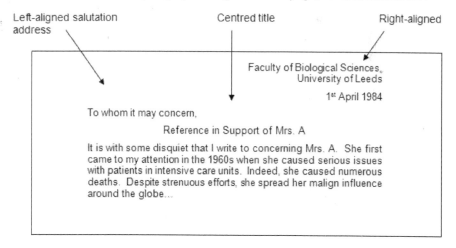

The paragraph is 'justified' where extra spaces have been added between words so that the right and left margins are neatly in line down the text. The margins for this can be set as appropriate for the page.

Tabs

Using appropriately chosen 'tab stop' positions can give a very polished look to your work. The default setting is for these to be regularly spaced across the page; each stop advances text by one unit[1] but you can alter them to suit your requirements either within the Paragraph dialogue box or directly on the ruler above the document. By clicking in the ruler, a bold 'L' appears. You can slide this backwards or forwards to set the first tab position. Clicking on the ruler again will give a second stop... and again a third stop and so on. This allows you to tabulate text like old-fashioned typists used to do.

On the top left of the editing window there is a box that by default has the 'L' symbol. Clicking on this allows you to scroll through other tab stop types. The decimal tab aligns numbers with the decimal point as the stop point. Why not experiment with the different tab stops to see what effect they can have?

Lists with bullets and numbers

These are extremely useful for presenting information clearly. You can choose the style of bullet to use and change the format of your numbered list to include letters. Multilevel options automatically create hanging indents as illustrated below.

- bullets
  - ➤ symbols (count as 'bulleted')
    1. numbers
       a) letters (count as 'numbered')

Borders and shading

Borders and shading can be applied to all or part of a selected paragraph to draw attention to specific text

Tables

Using a table is an economical and neat way of displaying many types of scientific data and also dispenses with the need to use repeated Tab spacings. You can easily organize the appearance of cells and the alignment of text within cells in the table, as illustrated below, using the Design and Layout menus that appear when the table is selected. To perform bulk operations on the whole table, you should click on the small blue cross that appears at the top left hand side of the table when you click anywhere in the table. This causes the whole table to be selected ready for formatting.

---

1 *Word* uses inches as the default setting for its ruler, harking back to the days of hot metal printing presses. If this seems old-fashioned to you, you can easily change the units to suit your preferences. Go to File > Options > Advanced > and scroll down to the 'Display' section. You can choose to use inches, centimetres, millimetres, points or picas.

Text may be positioned in different parts of a cell by changing the 'Cell Alignment'	In this cell, the text is aligned in the bottom right-hand corner...	The horizontal borders have been removed from the cells in this column (using 'Borders and Shading').	This cell has text aligned at the top and is centred. The cell below has had bolder borders added	
Individual cells can be shaded via 'Borders and Shading' in the shortcut menu.	...whereas in this one, it is aligned in the top right.		Cells can split into	also be subcells
Several cells may be merged (select cells, then choose 'Merge Cells'). The text has also been centred. This is useful for table titles and legends.			The text in this cell has been rotated through 90°	

Rows and columns may be added to or deleted from tables if you discover that you have planned for too few/many cells using the right click shortcut menu. If your table spreads over two pages, you can use the 'Split Table' button to spread the table properly; this allows you to copy the heading of the table from the top and paste it onto the top of the second page.

Word does distribute contents to provide the 'best fit' for data but you can alter the cell properties by, for example, dragging sliders in the ruler.

Converting text to tables and vice versa

It may be more convenient to deal with text within a table because you can format borders (or shading) more easily in the table format. Characters such as commas, paragraph marks, tabs or your own defined characters are used to indicate where the text should be divided into cells within a row and a paragraph return indicates where a new row should begin.

Sometimes data in a table are more usefully manipulated as plain text. Again, this is a simple process.

Defining your styles for a polished document

Word provides you with pre-set styles that define all the characteristics of the fonts, paragraphs, line spacing, headings, tabs, margins, etc. These are found in the 'Styles' menu. Using 'Styles' rather than formatting everything by hand allows you to produce a harmonized standard 'look' across the entire document, even one that is very large. If you prefer to create your own styles, this can easily be done and they can be saved as a 'Word Template' for use in future documents. Two particularly useful features for creating polished documents are listed below.

Tables of contents (and lists of tables and figures)

This is of particular value when producing large pieces of works with several different sections. You could use the default styles in *Word* but you will get much more control of the design of your table of contents by going to the 'Table of Contents...' link, which opens its dialogue box. Providing you use the heading styles consistently throughout your document, tables of contents such as that shown below (together with the type of heading used) give a neat and professional look.

Contents   (Heading 1)

Preface   (Heading 2)
  1 The microbiology of soil and of nutrient cycling   (Heading 3)
    1.1 What habitats are provided by soil?   (Heading 4)
    1.2 How are microbes involved in nutrient cycling?
      1.2.1 How is carbon cycled?   (Heading 5)
      1.2.2 How is nitrogen cycled?
      1.2.3 How is sulphur cycled?
  2 Plant–microbe interactions   (Heading 3)
    2.1 What are mycorrhizas?   (Heading 4)
    2.2 What sorts of symbioses do cyanobacteria form?
    2.3 What symbioses do other nitrogen-fixing bacteria form?
    ...
    8.8 What causes antibiotic resistance in bacteria?
Further reading...   (Heading 2)

It would have been a nightmare to compile that table of contents by hand! Lists of tables and figures can be created in a similar manner.

Cross-referencing

Again, with long documents it may be necessary to insert cross references to figures, tables and footnotes. By using the Styles facility, *Word* keeps track of the page where the item is positioned so that you don't have to worry about checking the page numbers if you reformat your document.

The page layout

By default, a new *Word* document opens such that every page is shown in 'Portrait' layout. Selecting the 'Landscape' option will convert the whole document to that page orientation. Sometimes, however, it is desirable to display specific sections of the material in 'Landscape' format, for instance, when displaying tables with a large number of columns, while retaining the rest in Portrait orientation.

Section Breaks

Introducing a 'Section Break' gives the freedom to format the selected section of the document in a different manner from the rest. After inserting a section break, you can adjust the page numbers, header and footer sections and any other type of formatting, such as increasing the numbers of columns of text, to what you wish in the new section. If you want different headers or footers in adjacent sections you must ensure that the 'Link to Previous' option is turned off. Conversely, make sure that it is turned on if you want continuous headers or footers.

Page Breaks

To avoid a heading appearing at the bottom of a page while the body of the text associated with that heading appears at the top of the next page, you can insert an appropriate Page Break.

Footnotes

When you are explaining complex ideas and you do not want to interrupt the flow of your text, it is a good idea to make use of footnotes. This can be done by accessing the 'Insert Footnote' facility.

Inserting figures

Figures enliven documents and illustrate important points in the most efficient manner. You can make your own figures, building up complex images from simple components; using the 'Shapes' menu, you can use the 'Clip Art' library to insert stock images and you can add your own image files from your computer. If an imported image requires a lot of storage space, you may need to reduce its size by saving it as different file type such as a '.jpeg', or '.gif'. Before doing this, check that the new file type is suitable for the image you want to display in terms of, for instance, the resolution and range of colour.

Creating a drawing using Word

Surprisingly complex figures can be produced using *Word*, as illustrated. You can build up the drawing using combinations of the set shapes, or draw 'freehand'. If your lines are slightly mis-positioned, there is an 'Edit Points' facility that helps you make adjustments. Molecular structures will require a mixture of text boxes and set shapes. It is wise to use the 'Select' tool followed by 'Group' facility to ensure that the various components do not drift apart if you have to reformat your document. It may be useful to 'Snap' components to each other or to a grid in order to get them appropriately aligned.

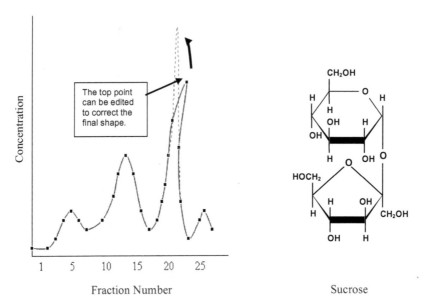

Figure A1 Illustrations of drawings in *Word*

Positioning the figure

Clicking on an image will show the 'handles' that you can use to resize your picture. By holding down the 'Shift' key when dragging the image to resize it from the corner of an image, you can ensure that the proportions of the picture are conserved. The 'Text Wrapping' facility gives a number of options for the positioning and presentation of pictures in a document.

Review options

In addition to providing help with spelling and grammar and providing a thesaurus, *Word* has additional features that enable you to keep track of any changes you might make to your own work or when working on group projects. Three very useful ones are described below.

Compare

If you have stored several versions of your work but are unsure which one is the latest, you can compare the different files to highlight any differences.

New Comment

When working on a particularly long document, such as your final year project, it may be useful to add a comment to a particular piece of text (via the 'New Comment' button) so that you can recall your thinking. This is also very useful when working as a team, since a comment can be used to explain to other members of the team why changes have been made or to pose questions that can provoke discussion among the team.

Track Changes

The 'Track Changes' option is of particular value when revising old documents. 'Track Changes' also helps to show which members of a team have contributed to the sub-editing of a multi-author document. When this function is turned on, contributions from each editor are shown in a different colour and hovering a cursor over edited text will reveal what changes have been made and who made them.

Perfecting *PowerPoint* presentations

PowerPoint has a number of sophisticated features that enable you to lift your presentation significantly. However, it is easy for enthusiastic novices to overdo the tricks, using all the 'bells and whistles' available. Do not get carried away with these, and always try to keep in mind what is appropriate for the audience and the context of your presentation. For example:

- try to keep slides simple and without clutter – it is harder for the audience to pick out the important points when there is too much on display;
- use one or two simple fonts that are large enough to make it easy to read (e.g. Arial or Calibri, and at least 18 point);
- use simple phrases rather than complex sentences with sub-clauses, so that audiences focus on the main points;
- if you want to include pictures, bear in mind that they may slow down the display;
- use a background that doesn't obscure the text or diagrams.

A fancy font, while it has a place in circus posters, is far too 'fussy' to be useful,

(EVEN WHEN YOU USE A LARGER FONT SIZE)

Standard formatting

When you open *PowerPoint*, you will be presented with a single slide that has two default areas for editing: a box that invites you to 'Click to add a title' and a second that invites you to 'Click to add a subtitle'. *PowerPoint* provides a variety of standard layouts but you can also design your own. Plain white slides are the default, but there are various 'themes' that you can access. Many of these default themes are rather on the bright side for formal presentations where the message is more important than the means of delivery, but they can be refined to suit your purposes. You can also insert your own background, which may be plain, graded or even a full-page image (providing it does not obscure your text).

On new slides, text is, by default, preceded by a bullet point. Individual bullet points may be indented. This changes the appearance of the text.

- This is how, by default, text appears on a slide: each paragraph forming a single 'bullet point'.
  - o 'first level' points often contain subordinate levels, starting with the 'second level',
    - then follows the 'third level',
      - o next the 'fourth level',
        - ➢ or even a 'fifth level', but by that stage, the default font size is getting hard to read.

At each change in level, the bullet style also changes and the text size gets proportionately smaller. Indenting a bullet point will change it to the next level down (and this can be reversed using the 'Decrease Indent' button). This allows you to highlight the logical structure of the slide. You can overwrite this if it does not suit your purposes. You can even suppress the use of bullet points if you wish. Text in the slides can be edited in a similar manner to that in *Word* and facilities such as the 'Find and Replace' can be used systematically throughout the slides.

Slide Master

This is a facility well worth investigating since it can save you a huge amount of time. If you do not want to use the standard templates provided by *PowerPoint*, the Slide Master allows you to create your own templates so you do not need to edit the style of each slide by hand. You can set the font size and formatting (bold, italic, type, colour, etc.) line spacing, positioning of the text boxes, and so on. It will ensure consistency throughout the slides, preventing, for instance, objects on hand-drawn slides appearing to jump on moving from one slide to the next. The Slide Master can also place page numbers, footers, logos and background images on all the slides automatically. Moreover, if you want to adjust the whole style of presentation for a different audience, you can change all the slides simultaneously simply by re-editing the Slide Master.

Building your slides

Although you can use the templates provided to make your presentation, you can also change the default attributes to suit your purposes. These can be manipulated easily, as in *Word* documents. A 'Blank' slide option is also available for you to add the content in your desired format. Things to bear in mind when creating your slides by hand are:

- **Text** can only be inserted within text boxes.
- **Notes** to describe the slide and aid your presentation can be added in the Presentation Pane. This is a pane that initially contains 'Click to add notes'. By clicking on the frame, you can add notes to describe the slide (you can drag the border up to allow you more room to view this pane and the presentation simultaneously shrinks. Notes can be useful when printing your presentation.

- **'Snap objects to grid'** can be a useful facility (particularly when creating a poster, as described below) to help organize and align different elements in the slide. Objects can also be 'snapped' to other objects to help them line up.
- **Background images** should be set with high transparency and sent behind any text boxes.

Hyperlinks

PowerPoint allows you to set up hyperlinks to web pages, to various pages within the current presentation, or to files on your computer (probably the least useful hyperlink function).

Links to web pages are very useful for allowing the presentation to include external resources and, while *PowerPoint* presentations are typically linear, by linking to pages within a presentation the presenter may take various pre-planned paths depending on choices made by the audience. Hyperlinks can be inserted as text, labels or as Action Buttons. These make for a truly interactive presentation.

Animations

Used properly, 'Animations' enliven your presentation; used badly they are a huge distraction. They provide several ways for text or images to appear or disappear, be emphasized, or moved around the screen. Combinations of effects can be used together and so there is a need to manage these. This is achieved using 'Custom Animation'. Each object and its animation are assigned a number and sequential animations are listed in a toolbar. If you want to delete an animation, it can be highlighted then removed. If the order of animations needs changing, then the animation in the toolbar can be dragged and dropped to its new location.

A good use of animations is to allow sequential bullet points to appear so that your presentation 'builds up', which prevents your audience from being distracted by information that has yet to be covered. Alternatively, you can always import an 'animated gif' file. These are image files where different layers have different images and, on display, the layers are flicked through to give the effect of a moving picture. With care, you can produce stunning effects with animations.

Slide transitions

PowerPoint provides a variety of visual effects for leading from one slide to the next during your presentation through the 'Transition to this Slide' facility. You can also set the speed and timing of the transition. It is a good idea to apply the same transition to all slides in your presentation, which can be done through the Slide Master. An exception may be made when your presentation is divided into discrete sections. In this case, changing transitions between sections will act to emphasize the different sections of your presentation.

Organizing your presentation

When reviewing your presentation you may decide that it would be better in a different order, or that you need to add a number of things that you have forgotten to include. This can easily be dealt with by using the 'Slide Sorter'. This shows all of the slides in your presentation as thumbnails, which can be dragged and dropped to different positions. To insert a new slide (or a duplicate) in the middle of your presentation, click on the space between the slides where you want to make your addition, go to the 'Home' tab and add a new slide or slides using the 'New Slide' button.

Printing handouts

PowerPoint has a number of 'Print' options. These open a dialogue box that allows you to make a number of important choices, as detailed below:

- **'Handouts'** where you can pick how many slides per page you want to print from the given options of 1, 2, 3, 4, 6 or 9 slides per page. The default is six slides per page and you can set the order in which they appear. If you elect to print nine slides per page, the resulting hard copy tends to look cluttered. When giving talks where you make asides and add additional information, the three-slide option is useful since the print includes space for your audience to add their own notes

- **'Slides'** which allows you to print off individual slides; however, this is very wasteful unless you really do need whole page reproductions

- **'Notes Pages'** is useful if you have annotated slides and want to use these to help you with your presentation, especially when little or no text appears on the slide. Be aware, however, that with this choice each slide will have its own page.

- **'Outline'** where only the text that appears in the title box of a slide is printed.

It is ALWAYS worth checking the final layout using the 'Print Preview' option before committing to print.

Posters

Posters can be prepared in *PowerPoint* using a single slide. They can be built up using all of the components that may be used for other presentations: text boxes, images, charts and tables, and so on. Due to their large size, specialist printers must be used to produce paper copies of the finished posters. Standard poster sizes are shown below.

Table A2 Paper sizes and dimensions.

Size	Dimensions (cm)
A0	84.1 x 118.9
A1	59.4 x 84.1
A2	42.0 x 59.4
A3	29.7 x 42.0

The main points to bear in mind when creating a poster are:

- **Is there a required orientation?** Before starting your poster it is essential that you check whether you must present it in 'Portrait' orientation or (the default) 'Landscape'
- **Is there a size restriction?** Again, you must check.
- **Leave room for a margin** on your printed poster, as printers often do not print right to the edge.
- **Align and arrange objects** using the 'Grid Settings' facilities. It can be tricky aligning objects accurately, particularly with very large posters, where what look like small inaccuracies on screen can blow up into massive issues in your A0 poster. To help to get alignments right, you should use 'Snap objects to grid' and 'Snap objects to other objects'.
- **Will the images retain their clarity as they are enlarged?** Since the poster will be large, images that look fine on your PC monitor or printed on an A4-sized sheet may become indistinct once blown up to the final size, so do set the poster size before you do anything else.

Both the size and orientation of the presentation can be manipulated using the 'Page Setup' and 'Slide Orientation' boxes in the 'Page Setup' box in the 'Design' facility.

Excelling in *Excel*

Excel is a spreadsheet that allows you to analyse and manipulate data in numerical and text-based forms and even in symbols. In addition to straightforward ordering by numerical rank and alphabet, data can also be analysed using functions embedded in *Excel*. This allows you to perform statistical analysis of your data with ease. *Excel* also allows you to produce a number of different charts to display your data. Used to its full potential, *Excel* is an indispensable tool for data analysis and presentation.

Workbooks, worksheets and cells

When running *Excel*, you are presented with a workbook comprising three worksheets initially. By default, the sheets in a workbook are named '*Sheet 1*', '*Sheet 2*' and '*Sheet 3*', respectively; their names appear at the bottom left of the worksheet. You can change the name of sheets, add or delete them by right clicking on the tab to open a dialogue box. Having several worksheets within one workbook allows you to copy data from one to another relatively easily and also means that when you save a workbook, all the spreadsheets contained within that workbook will be saved into the same file.

Each worksheet comprises columns, labelled with letters and rows, labelled with numbers so it is easy to identify each cell (where a row intersects a column) by a unique reference, comprising the column letter(s) and the row number.

You can only manipulate data in 'active' cells, which are highlighted. Once highlighted, the reference of the cell appears in the 'Name' box at the top left of the worksheet. Using the arrow keys, you can move around the cells of the worksheet, with the highlighted cell becoming the 'active cell'. Alternatively, you can move to an active cell by typing its name in the name box or by moving the cursor over the cell and then clicking.

When working with a very large spreadsheet that does not fit entirely onto the screen, it is often convenient to 'Split' the screen. This causes the spreadsheet to become divided into two vertical and two horizontal portions, with the ability to scroll along each portion independently of the others. If you want to stop a selected portion of the spreadsheet from moving completely, for example, the top row or first column, you can do this using the options available in 'Freeze Panes'.

It is worth noting that the cursor changes appearance according to what function it is about to perform. Normally it will be a chunky cross, but when hovering over the bottom right-hand corner of an active cell it becomes a black cross, indicating that you will be able to drag your cursor over a number of different cells in order to select them for a bulk operation. Hovering over any other corner will convert it into an arrowed cross. This means you can perform a drag and drop action on the contents of that single cell.

Entering data

To enter data, select the cell in which you want the data to appear and type the value. What you type appears in the cell and also in the formula bar. If you want to edit the cell, you can either type directly into the active cell or into the formula bar. Pressing enter will move the active cell one cell down the column in which you are working. To ensure your data are entered in the correct position, it is a good idea to get into the habit of labelling your rows and/or columns first.

Replicating data

A way of replicating data is to enter your data in the active cell, click on the box at the bottom right of the active cell and drag this across to the right to replicate the data in the columns to the right or drag the box down to fill rows underneath the active cell. When using this technique, dates are preserved rather than being increased. If, however, you have two consecutive dates, or names of months, or names of the days of the week, etc. in adjacent cells, then by selecting both cells and dragging the box, adjoining cells will be filled with incremental dates, months, days and so on.

Transferring data to and from Word

Data may be easily transferred from (or to) a *Word* document. Copying the contents of a table from *Word* to the clipboard allows you to paste the data into an *Excel* worksheet. Simply place your cursor in the cell that you want to be the top left cell of the data and paste the contents of the clipboard. **Note**: ensure there is sufficient 'free' space in your spreadsheet for the pasted material because any existing data within the pasted area will be overwritten.

Transferring data from the web

You can also transfer data from web pages in a similar manner by highlighting then copying the relevant material to the spreadsheet. It is worth pasting this into a new *Word* document before transferring to *Excel* because it is easier to 'tidy' your data in *Word*, removing unwanted tags and characters, than in *Excel*.

Formatting your data

To avoid text appearing truncated when the column to the right has data entered, the 'Format Cell...' option provides access to a dialogue box with a number of tabs. The 'Alignment' tab is good for controlling how long strings of text appear. Especially useful is the 'Wrap texts' function.

Numerical data can also be formatted in numerous ways, for example the 'Number' tab on the dialogue box for formatting cells. You can control the number of decimal points for plain numbers, or choose to display data in a currency format or as percentages. Data can be converted to times or dates in various formats, and so on.

Working with formulae

As well as text or numbers, cells can also hold formulae. It is the application of formulae in order to perform calculations that gives *Excel* spreadsheets their power. Since cell locations are specified rather than the values that they contain, if you change the numerical value within a cell that is specified in a formula, the result is automatically recalculated. All the formulae start with an equals sign '=' so *Excel* knows that a formula is being entered rather than just text.

Note: if you do actually want to enter text preceded by '=', you must type an apostrophe before the '=' to let *Excel* know it is not a formula.

Excel will automatically put the calculated result adjacent to the last cell in the row or column of data. You can, however, drag this (together with its associated formula) to another cell once you have selected the cell and the cursor turns into the little 'arrowed' cross.

Commonly used calculations

These can be accessed through the 'AutoSum', which inserts the relevant formulae to a set of cells that you have selected (by dragging the cursor over them and/ or individually clicking on them). The 'More Functions' option provides additional formulae but if the one you want is not visible, you can search for it. You can check that all the cells you want to analyse are included by looking at the range specific in the formula bar, and also in the 'Function Arguments' box that appears when you use the 'More Functions' facility. If you want to create your own functions, the table below illustrates the basic notation used by *Excel*.

Table A3 Common functions.

Symbol	Example	Function
+	=(B2+F3+25)	Adds
	=(B4D25)	Subtracts
/	=(D9/A2)/100	Divides
*	=(B7*C3)	Multiplies (remember not to use 'x', which is just a letter not a mathematical symbol in *Excel*)
SUM	=SUM(A2:A25)	Adds the range of cells specified (in this example, all the cells from A2 to A25)
AVERAGE	=AVERAGE(A2:A25)	Calculates the mean value of the range of cells specified (A2 to A25)
STDEV	=STDEV(A2:A25)	Calculates the standard deviation of the range of cells specified (A2 to A25)
^	=A5^3	Raises the value specified to the power of whatever number appears to the right of the symbol (in this example, A5 would be cubed)

The 'if' function

This is very helpful for comparison of two sets of data that may be numerical or textual. The syntax is:

=IF([Logical test], [value if true], [value if false])

Consider the function =IF(A1>A2, 1,0), located in cell A3. When applied to two adjacent cells A1 and A2 it will return a value of 1 in column A3 if the data in cell A1 is greater than that in cell A2 and a value of 0 if the value of the data in cell A1 is the same as or less than the value of the value of data in cell A2. For two long columns of data, it is very easy to scan down the result column and draw your conclusion. Text data may also be compared, so the function: =IF(A1='precipitation', 'Yes', 'No'), will return a value of Yes if cell A1 contains the text precipitation and the value of No if it does not. The comparison is not case dependent and a value of Yes will be returned for precipitation, Precipitation and even PrEcIpItAtIoN but will return No if cell A1 contains precipitations.

Fixing the reference cell

Typically, cell references are 'relative' so that if the formula in a cell is copied or cut and pasted to another location, the cell references will move to cells that are placed in the same relative position to those in the original location. This means that if you were to copy an active cell with reference to cell A1 and paste it into a cell three columns to the right, your analysis would then refer to data in cell D1 rather than that in cell A1 and your formula is likely to return a spurious result from using inappropriate data. In these circumstances, you can fix the absolute column reference and/or the row reference by adding dollar signs to the reference, thus $A will mean that the column reference is fixed to Column A. Likewise $1 fixes the row reference to Row 1. Placing a dollar sign before both the letter and the number in a cell reference will fix both the column and the row position, thus C12 will always refer to the cell in row 12 of the third column of a spreadsheet, no matter where its formula gets copied.

Error messages

If there are problems with the formula you have entered, or some of the data doesn't fit appropriately into the formula, *Excel* will produce an error message. If the message appears due to an impossible result, you can insert an asterisk in that one cell, and then proceed with the others as normal. Below is a table describing the common errors people encounter when using *Excel*.

Table A4 Common error messages.

Error code	Meaning
####	The cell is too narrow to display the contents – make the column wider to fix this issue
#DIV/0	The formula is trying to divide by zero – check that the formula is correct and that you are not instructing the formula to divide by cells containing zero (in this context, blank cells are treated as zero by *Excel*)
#NAME	*Excel* doesn't recognize a name used in the formula – check your typing
#NUM	There is a problem with a number in your formula – check the numbers
#REF	The formula uses a cell reference that is not valid – check that you have specified a suitable range of cells and that their names are correct
#VALUE	You have used the wrong type of number or have attempted to calculate using a cell containing text – check your typing and the specific cells

Saving time with calculations – spreading formulae

Excel has a very useful 'Autofill' function that allows you to 'spread' a formula to adjacent cells. This means that a calculation performed using one set of cells can be automatically applied to a range of others without you needing to type in formulae for each successive set of data. For example, if you want the mean values of each of several rows of data, you can perform the function on the first row, putting the result in a cell at the end and then 'spread' the formula down the column. Make sure you check in the Formula bar that the cell range suggested by *Excel* is correct each time.

This 'spreading' method can also be used for filling in successive numbers, dates or times. For example, if you want to number a long column of items, put '1' in the cell to the left of the first item, '2' in the cell below that, then select these two cells, click on the 'cross' icon and drag down the column: successive cells will be filled with '3', '4', '5', etc. Additionally, if you type 'Monday' into a cell, spreading the 'formula' will cause successive cells to be filled with 'Tuesday', 'Wednesday', 'Thursday' and so on.

Transferring calculated data within and between spreadsheets

As detailed above, when formulae are copied (or cut) and pasted from one cell to another, they are automatically adjusted to operate on corresponding cells. So, for example, when a formula in cell F2 that gives the sum of cells A2 to D2 is copied to cell F8, the numerical value in F8 will be the sum of cells A8 to D8. This means that you cannot just copy columns of calculated values to another part of the spreadsheet or to a different spreadsheet, since those values will change. *Excel* provides a solution to this problem through the 'Paste Special' facility. Having copied (or cut) the relevant cells to the Clipboard, you should use the 'Paste Special/Values' (or choose 'Paste Special' from the shortcut menu). Alternatively, you can paste the calculated data first into a *Word* document (which doesn't attach the formulae to the numbers) and then transfer it all back into a new *Excel* worksheet.

Sorting data

Sorting your data appropriately allows you to analyse data and reveal patterns much more easily than would otherwise be possible. When sorting a whole worksheet it is quickest to select it all by clicking the box in its top left-hand corner.

The 'Sort' function provides different ways of arranging the information and allows for extra levels of organization within the initial order. If you have selected the 'My data has headers' option, *Excel* will use the column names in the sort menu; otherwise it will simply give the letter heading the column.

The 'Filter' option allows you to deselect any data you temporarily wish to ignore. At the head of each column of data dropdown menus appear, from which you can deselect data. Filters in different columns can easily be combined. Proof that the data have been filtered rather than edited out can be seen by gaps in their corresponding row numbers.

Pivot tables provide a further way of sorting data. They 'pivot' data from one worksheet to another in order to combine and filter data in different ways, simply by dragging and dropping data fields from one part of the 'Pivot Table Field List' panel to another. Unlike the 'Filter' facility, the resulting worksheet contains no gaps.

Charts

When analysing data, it is often more useful to view the data graphically. Simply select the range of cells that you want to view, including the data labels, and click on the 'Insert' tab, where you will find a range of standard chart types (others may be accessed through the 'Other Charts' button).

Presentation

To give the best possible appearance you can easily alter most aspects of your chart, including labelling the chart and its axes, redefining the range of data that are displayed and altering the range of the axes to spread the data more evenly along the chart. You can also switch the horizontal and vertical axes.

Error bars

Any measurement is subject to error. These are conventionally shown on charts by including 'error bars'. Although clicking on the 'Standard Error' option will apply bars to the tops of the columns, you may need to modify these. By default, *Excel* adds error bars derived from the whole data set. With real data, this is not a reasonable assumption. Fortunately, *Excel* allows error bars to be added for individual datum points through its 'Custom' option.

Trendlines

If you want to see the 'line of best fit' for your data, *Excel* can insert a variety of different types of trendlines, including linear, exponential, logarithmic and moving averages, through the 'More Options' facility in the 'Add Trendline' option. But that is not all that you can do. You are given the option to include the equation defining your line (in the form of y = m x + C) and the correlation coefficient 'R^2'. This is a measure of the degree to which your data diverge from the line of best fit where the closer is its value to 1, the closer are the observed data to the line of best fit.

Becoming adept with *Access*

Excel spreadsheets can manage large amounts of data easily but in computer jargon, the data in spreadsheets are described as 'flat'. Sometimes you may want to relate several types of data together from different spreadsheets. This is where databases such as *Access* come in. *Access* is called a **relational database** because it lifts and organizes the 'flat' data from individual tables by relating information between them. Although it may appear a little intimidating at first, *Access* is an extremely powerful tool that will allow you to manage vast amounts of data. To use *Access* you must be familiar with the following terms:

- **Tables**: Individual *Access* tables are a bit like single *Excel* spreadsheets, in which each row holds a 'record' (e.g. details of one group member), and each column contains a particular 'field' (e.g. forename, surname, postcode, phone number). Fields can take common data types such as numbers, text, dates, currency and even images. *Access* enables you to define what data are entered, which means that only the correct data type can be entered in any given 'field'. For example, you will not be able to enter a telephone number in a field designed to store email addresses.

- **Primary Key**: The primary key in an *Access* table is a field – for example a number – which is used to 'index' the data in the table, and must be unique for each record (i.e. no two records can have the same primary key). By including the primary key from one table as a field in another table you can make relationships between different tables. While *Access* can set its own primary key for each table using the 'Autonumber' feature, you can also choose your own.

- **Queries**: Queries are powerful tools that allow you to sort or filter your data so that only particular types of information are retrieved (e.g. surname and phone number for all group members with a particular postcode). Searching for information that may be embedded somewhere in a very large table would be a nightmare without this facility. Queries can be saved so that you can re-use them in the future for similar searches.

- **Forms**: These allow you to combine information from multiple tables and can be used to display the results in a user-friendly interface to allow you to work easily with large data sets. Data can be displayed in a number of different ways. As well as using windows to display the contents of a particular field, you can use drop-down menus to allow users to pick from a list of options. Buttons can also be built into forms to bring about actions, such as allowing users to open other forms in the database. Designing a good form will not only make your work more efficient, it will also allow you to share your database with other people. Furthermore, you can secure fields within a form so that other users can only modify the data that you wish them to be able to change. This makes *Access* an ideal tool for teamwork.

So, how can all of this be applied?

Imagine that you are a bacteriologist interested in the spread of antibiotic resistance in a hospital environment. Antibiotic resistance in bacteria is frequently encoded by mobile genetic elements that can move from one bacterial strain to another. Sometimes they can transfer to bacteria of different species or genera and occasionally to bacteria in different families. Because these genes are mobile, bacteria may acquire resistance to more than one type of antibiotic. Tracking this can be quite a task.

Using *Access*, you could first build your *isolates* **Table**, with all of the bacteriology information in a single spreadsheet. For example, you could create a spreadsheet where the **Primary Key** is the unique isolate number generated by the lab. You will want to include other information concerning each isolate: its identity in terms of genus and species, the date of isolation, the type of specimen from which it was recovered, typing information to allow strain discrimination, and its susceptibility or resistance to a range of relevant antibiotics. If you then want to discover how resistance is spreading through bacteria in your hospital, to help you formulate ideas about how to contain the problem, you will need further data, including information about the people from whom the isolates were recovered.

You could generate a second table with *patient* information data. The **Primary Key** in this case would probably be the hospital number. In this table, in addition to each hospital number and the name of each patient, you could store information regarding the age and gender of the patient and any underlying medical or surgical condition, and so on. Of particular interest in this table will be what bacteriology specimens have been submitted to the lab for each patient.

Further tables that may be helpful could hold information on medical, paramedical and nursing staff. Entering data in each table will be much easier when you design purpose-built **Forms**; and once you have built the relationships between the tables, you can generate **Queries**. If you have designed a really good database, you may be able to trace the occurrence of a particular gene as it spreads from one bacterium to another, and to trace the movement of bacteria carrying resistances between people and also between hospital units.

From this example, it should be apparent that even when applying the 'Filter' function, a single spreadsheet in *Excel* would be unable to link such diverse and complicated data, nor be able to retrieve information in such a selective manner.

Note: You should bear in mind that the scenario described above means storing sensitive personal details. If working on such a problem for real, you would certainly have to obtain ethical approval and provide the local ethics committee with details of how you will ensure the security of your data. Fortunately, *Access* allows you to encrypt your data and to protect your databases with passwords that you can set.

- **Reports**: These generate selective information derived from the *Access* database for 'export' to people who are not using your database. Reports may be generated from a single table, or from a number of tables in the database, and may give rise to physical copies (when printed, for example), or can be sent as email attachments, or may be delivered as part of a web page when the information is made available to a web server.

Planning your database

Successful databases require careful planning so that you can store, analyse and retrieve data efficiently. You **must** have a clear idea of what you want from your database **before** you start to enter your data. This means thinking about all the information you may need to store in your database (write it all down) and how you can group information together so that all of the data of a particular type are in a single table.

Without a clear plan, your database will quickly descend into chaos!

In formulating your plan, think about:

- the types of data that you want to work with;
- how your data sets relate to each other;
- whether you have considered all of the relevant variables that you need to include in your database;
- the best way to enter data;
- how best to use the data.

As you become confident with *Access*, you can explore 'macros' (simple computer programs), to help with repetitive tasks, and you can go on to use the Visual Basic programming language in *Access* modules. For most users, however, these advanced tools are not needed and you can create great databases simply by sticking with the basic functions. *Microsoft* provides online training for *Access*, but its emphasis is on business applications. Do not let that put you off. Other training materials are available. Although it may appear a little intimidating at first, *Access* is a great tool, allowing you to store and analyse complex data with relative ease. Take the plunge – give it a go! Making a very simple database with contact details of your mates is not a bad place to start...

Index